Special Monograph

Online Instruction For Distance Education Delivery: Preparing Special Educators In and For Rural Areas

Barbara L. Ludlow
West Virginia University

Belva C. Collins
University of Kentucky

Ronda Menlove
Utah State University

American Council on Rural Special Education

© Copyright 2006 Barbara L. Ludlow, Belva C. Collins, and Ronda R. Menlove, American Council on Rural Special Education.

All rights reserved. No part of this publication may be reproduced, stored in a retrieval system, or transmitted, in any form or by any means, electronic, mechanical, photocopying, recording, or otherwise, without the written prior permission of the author.

Front cover photo: Mary Susan Fishbaugh

Note for Librarians: A cataloguing record for this book is available from Library and Archives Canada at www.collectionscanada.ca/amicus/index-e.html
ISBN 1-4120-8238-2

TRAFFORD
PUBLISHING

Offices in Canada, USA, Ireland and UK

Book sales for North America and international:
Trafford Publishing, 6E–2333 Government St.,
Victoria, BC V8T 4P4 CANADA
phone 250 383 6864 (toll-free 1 888 232 4444)
fax 250 383 6804; email to orders@trafford.com

Book sales in Europe:
Trafford Publishing (UK) Limited, 9 Park End Street, 2nd Floor
Oxford, UK OX1 1HH UNITED KINGDOM
phone 44 (0)1865 722 113 (local rate 0845 230 9601)
facsimile 44 (0)1865 722 868; info.uk@trafford.com

Order online at:
trafford.com/05-3204

10 9 8 7 6 5 4 3 2 1

Preface

Kay S. Bull
Oklahoma State University

This book presents methods, processes, and examples of online learning for special education course developers in rural and remote areas where many teachers teach special education with emergency certificates. To rectify this situation, courses and programs have been developed by a number of universities. In this book, you will see example courses and programs developed specifically for use. The first three chapters in the book deal with issues in developing online materials that can be constructed to create courses for rural special education teachers (RSETs). These chapters provide direction on various kinds of development that are basic to online course creation, such as how to do it (Ludlow – Chapter 1), planning and delivering (Collins & Gaylon-Keramidas – Chapter 2), and providing technological support (Mitchem – Chapter 3).

I was asked to develop this preface because I have taught instructional design for the past 23 years and, in 1996, I began teaching online for that reason. I was the guinea pig for the College of Education at Oklahoma State University and taught its first totally online course. It was a discussion-based format, one I carried over from the traditional classroom and taught on a Lotus Learning Space discussion board. Everyone talked to me and to their peers, and I talked to everyone. We all talked and discussed the course material; this was very different from the traditional classroom in which a minority discuss while the majority remains silent, but it is the same finding reported by Knapezyk in Chapter 6. In online courses, everyone has to talk. This experience made a convert out of me. I had never had a course where everyone talked every week, even shy female international students who rarely speak up in the traditional classroom. The diversity of response was amazing.

Course, content, cultural, ethnic and perspectival issues were raised that I had not heard in 19 years of teaching. It was a wonderful experience. In 1997, I started converting courses to online asynchronous experiences, and I have not taught a course in the classroom since. This kind of development of the discussion process also is described by Sebastian, Schwartz, and Duckett in Chapter 13. Discussion, however, should not grow too large. One thing that we have discovered is that there are few possibilities of economy of scale (teaching large numbers of students) online. The maximum number of students who can read and respond to questions, while reading the responses of others, is limited. With more than 25 students, the students will not take the time to read others' work, and this degrades their performance and their understanding of the content. Communities of learning fail to form.

Programs at a distance can be delivered to sites (see Agran & Kiefer-O'Donnell – Chapter 7 and Pemberton, Tyler-Wood, & Restine – Chapter 10) or to individuals. Both approaches work. Sites, however, may be quite distant from the locations of individuals. For this reason, in remote rural regions,

Preface

9) as examples. Students from all over the world may want to sign up for courses if they are well advertised. If the plan is for students to meet face-to-face or to do group projects, you will need to limit the area from which students will be drawn. (It is hard to schedule meetings if you are nine time zones away.) However, see the procedures described by Ludlow and Duff in Chapter 9 for using telephones with speakers for meetings. Those who teach RSETs probably should be willing to take on all comers, as the need is great in rural and remote areas.

A number of those who write about teaching in rural areas believe that RSETs will have access to wide bandwidth. This is probably true in eastern states, but that level of bandwidth has not yet reached the mid and far west; the distances are too great. Overseas is even a greater problem. Before you plan to send courses to RSETs in rural areas, check to see if they can use the sophisticated technology (e.g., video or synchronous presentations) you want to provide. Sophisticated technology can increase frustration if it does not work well on a 28.8 kbps phone line. Many presentations with video and audio are less than effective when the bandwidth is narrow. Chatrooms work well only in very small groups or when the teacher rules with an iron hand to control who talks and when. DVDs and CD-ROMs exceed the capacity of at least some computers at home, at school, and in Carnegie libraries. To reach RSETS in rural and remote areas, low tech is the best tech. Even schools and Carnegie Libraries in many small towns in midwestern and western states use 28.8 kbps bandwidth.

Once you have the course developed and students recruited, you will need to teach your students how to participate in courses. Teaching students to work online is not difficult if you are using the right platform (see Sebastian et al., Chapter 13). Training can be provided online, with paper instructions, or on a CD-ROM/DVD. Anything that moves, talks, twitches or wiggles should be presented to students on a CD-ROM or DVD. This circumvents the need to have a high-speed modem. RSETs are likely to live in low technology areas. Projects and course work that require high tech use should be confined to CDs or DVDs. Of course, this does not work for synchronous delivery. A CD at the beginning of the course is also useful for having all the instructions, plugins, etc., that are needed for the course burned to it. This is discussed by Steinweg and Warren in Chapter 5. My preference is for self-starting CD-ROMs, as you can make them for less than a dollar and mail them to rural, remote students. A half page of directions with the CD provides everything that is needed to boot it up. Computers that have CD-ROM drives can be required for participation in the course. For old computers, external drives can be bought and connected cheaply.

There are a number of formats for courses presented by the authors of this volume. Each of them is worth considering. As an old instructional designer who does everything asynchronously, I was intrigued by Knapczyk's ideas of "starters" and "wrappers," which you will read about in Chapter 6. Could I adapt this approach? As we develop new approaches to course design, our students will become better and better learners. We must continue to develop new approaches that will serve our students better in the future.

REFERENCES

World Wide Web Consortium. (1999) Web content accessibility guidelines 1.0 Retrieved August 30, 2005 from http://www.w3.org/TR/WAI-WEBCONTENT/#toc

Section 508. The Center for Information Technology Accommodation (CITA), in the U.S. General Services Administration's Office of Government wide Policy. Retrieved August 30, 2005 from http://www.section508.gov/

Contents

Preface   KAY S. BULL, *Oklahoma State University*............iii

PART 1   **DELIVERY OF ONLINE INSTRUCTION**

Chapter 1   Overview of Online Instruction................... 2
BARBARA L. LUDLOW, *West Virginia University*

Chapter 2   Planning and Delivering Online Instruction 43
BELVA C. COLLINS, *University of Kentucky*
CATHY GALYON-KERAMIDAS, *West Virginia University*

Chapter 3   Providing Technology Support to Learners 76
KATHERINE J. MITCHEM,
California University of Pennsylvania

PART 2   **PROGRAM DESCRIPTIONS**

Chapter 4   Online Delivery of Programs in Low-Incidence Disabilities and Special Education Administration to Meet Statewide and National Needs 98
HARVEY A. RUDE, *University of Northern Colorado*
KAY ALICYN FERRELL, *National Center on Low Incidence Disabilities*

Chapter 5   Online Master's Degree Program in Special Education at East Carolina University........... 112
SUE B. STEINWEG AND SANDRA H. WARREN,
East Carolina University

Chapter 6   Indiana University's Collaborative Teacher Education Program 122
Dennis Knapczyk, Indiana University

Chapter 7   Desktop Videoconferencing in Utah: The UPLIFT Project.......................... 136
Martin Agran, University of Northern Iowa
Richard Keifer-O'Donnell, Minnesota University

Chapter 8   An Online Special Education Alternative Licensure Program in New Mexico 151
Teresa Rowlison, New Mexico State University

Chapter 9   Live Webcasting: An Interactive Online Program to Prepare Personnel for Rural Areas 165
Barbara L. Ludlow, West Virginia University
Michael C. Duff, Discover Video Productions

Chapter 10   Desktop Videoconferencing for Practicum Supervision in Texas: Two-way Interactive Video and Audio with Polycom ViewStation 128 176
Jane B. Pemberton, Texas Woman's University
Tandra Tyler-Wood, University of North Texas
L. Nan Restone, Texas Woman's University

Chapter 11   Project IMPACT*NET: A Distance Inservice Model to Increase Classroom Skills of Paraeducators and Their Supervising Teachers..................... 188
David E. Forbush, Robert L. Morgan,
Utah State University

Chapter 12   The Professional Development School without Walls in Southeast Florida..................... 201
Betty C. Epanchin, Elizabeth Doone,
Karen Colucci, University of South Florida

Chapter 13 Supporting Master of Science in Special
Education Students Online: Challenges and
Tentative Solutions 215
JOAN P. SEBASTIAN, STUART SCHWARTZ,
JANE DUCKETT, *National University*

Chapter 14 Online Modules to Prepare Distance Educators
at the University of Kentucky 230
BELVA C. COLLINS, CONSTANCE M. BAIRD,
University of Kentucky

Summary CLEBORNE MADDOX,
University of Nevada Reno 249

Part 1

ONLINE INSTRUCTION FOR DISTANCE EDUCATION DELIVERY: PREPARING SPECIAL EDUCATORS IN AND FOR RURAL AREAS

Chapter 1

Overview of Online Instruction

Barbara L. Ludlow
West Virginia University

Online instruction is the most recent and fastest growing distance education delivery system. When the Internet became available for general use in the 1980s, colleges and universities soon saw its potential to replace burdensome mechanisms of correspondence study and expensive models of televised instruction with a new form of teaching and learning (Kearsley, 2001). These early Internet-based courses often relied solely on text documents and primarily asynchronous interactions through electronic mail and discussion boards. With the introduction of the World Wide Web and its multimedia capabilities in the late 1990s, online instruction evolved to take advantage of non-text media such as images, animations, audio and video and more synchronous interactions through text chats, instant messaging, and desktop conferencing (Moore & Kearsley, 2005). The current prediction is that 100% of higher education courses will utilize the Internet for instruction, either as a support for campus-based instruction or as the primary instructional delivery system by 2010 (Boettcher, 2004).

Educators were quick to recognize the potential of online instruction for personnel preparation in special education for both campus support and distance education applications. During the last decade, institutions of higher education have explored the use of online instruction to increase access to preservice and inservice training to rural schools and communities. This chapter presents an overview of online instruction and its components, reviews the literature on online courses and programs in special education, and examines current issues and emerging trends in using online instruction.

WHAT IS ONLINE INSTRUCTION?

Online instruction is the use of the Internet (text only) and World Wide Web (multiple media) for teaching and learning. Allen and Seaman (2004) present a continuum of instruction from traditional (no online component) through web-facilitated or web-enhanced (1-29% online) and blended or hybrid (30-79% online) to fully online (80% or more with no face-to-face meetings). Although web-enhanced instruction can be based on simple applications such as email or course Web sites, blended and fully online instruction must incorporate multiple applications. Online instruction can be used to support or extend face-to-face courses and workshops, to allow any time, any place access to instructional materials on campus (sometimes termed distributed education), or to deliver distance education programs to learners at remote locations.

Overview of Online Instruction

Worldwide access to the Internet and Web has grown by leaps and bounds in recent years. Where once only governments, universities, and large corporations had access to powerful computers and high-speed connections, today most schools and workplaces and many private homes have such access, even in quite remote locations. Both desktop and laptop computers come equipped with one or more Web browser programs (such as Microsoft Internet Explorer, Netscape Communicator, or Apple Safari), software applications that open and read content posted online. Users may use dialup phone modems at connection speeds of 28 or 56 kbps (although actual speed in some rural areas may be lower), or choose from among digital subscriber lines, T1 fiber cable, and cable and satellite television modems for connection speeds of 128 kpbs and up.

Delivery of online instruction requires a Web server (a computer with special software), a network (access to Internet with a domain name to allow users to contact the server), and one or more programs that support content that can be transmitted to client computers as packets of information via Internet Protocol (IP) (Ko & Rossen, 2004). Most institutions of higher education involved in online course or program delivery use one of a number of commercially available Web course management systems (CMS), which are collections of tools for online teaching and learning. The most popular systems include WebCT Campus Edition and Vista (http://www.webct.com), Blackboard (http://www.blackboard.com), eCollege (http://www.ecollege.com), and First Class (http://www.softarc.com). Each CMS offers a variety of tools (Dabbagh & Bannan-Ritland, 2005): content creation and delivery tools (modules, hyperlinks, document import and storage); collaboration and communication tools (mail, discussion boards, group project tools); administration and management tools (announcements, calendars, group generators, security and access settings); individual learning tools (folders, bookmarks, note taking devices, navigation aids, and search and help functions); and assessment tools (quizzes, assignment drop-boxes, grade books, and tracking and data compilation). Each CMS varies in the range of tools included and the learning curve required to master and use them.

The development of online learning materials and activities also requires the use of authoring tools (Dabbagh & Bannan-Ritland, 2005). Web design programs include Macromedia Dreamweaver (http://www.macromedia.com/software/dreamweaver) to prepare documents in hypertext markup language (HTML); Adobe Photoshop to prepare images (http://www.adobe.com/photoshop); Adobe Acrobat (http://www.adobe.com/acrobat) to prepare Portable Document Files (PDFs) that combine print and media; Macromedia Flash (http://www.macromedia.com/software/flash) to prepare animations and interactive activities, any of several media editing programs (for example, Microsoft Windows Moviemaker (http://www.microsoft.com/moviemaker), Adobe Premiere (http://www.adobe.com/premiere), or Apple FinalCut Pro (http://www.apple.com/finalcutpro) to edit and digitize audio and video media; and any of several streaming media programs (for example, RealNetwork RealProducer (http://www.realnetworks.com/products/producer), Microsoft Producer (http://www.microsoft.com/windows/windowsproducer), or Apple QuickTime Pro (http://www.apple.com/quicktime) to prepare streaming media files.

Pros and Cons of Online Instruction

Online instruction opens up new possibilities for instruction, but it also presents some constraints. Kearsley (2000) summed up the benefits of online instruction as economies of scale by increasing enrollments, increased access to new populations of learners, and enhanced quality of teaching and learning. He identified the disadvantages of online instruction as high development costs, ongoing costs associated with technical infrastructure (equipment and personnel), and incremental costs associated with increasing enrollment. Nevertheless, the twin forces of learner demand for greater access to higher education and the potential for lower costs due to economies of scale are driving colleges and universities to develop more online instruction to expand their course and program delivery options (Bates, 2000).

Two important considerations in designing any distance delivery system are access to the technology and quality of the media (Moore & Kearsley, 2005). To implement online instruction, not only must instructors and learners have equipment (computer and connection) that allows them to access the Web, but the program also must use media (text, still and moving images, sounds and artifacts) that the technology can distribute efficiently. At present, the online environment supports text and voice very well, simple still images and animations adequately, and music and video to some extent, although clarity and quality may vary depending on connection speed and processor speed. When Internet2 becomes more widely available, its improved network and greater bandwidth will facilitate delivery of real time audio and media of high technical quality, changing the face of online instruction yet again (Van Dusen, 2000).

WHAT ONLINE INSTRUCTIONAL FORMATS ARE AVAILABLE?

The online learning environment can be used in a variety of ways for content presentation and learner interaction during instruction. Online instructional formats may rely on asynchronous (delayed time) or synchronous (real time) formats (Kearsley, 2000). Asynchronous formats allow learners to access information or communicate with the instructor and other learners at their own convenience, while synchronous formats require all learners to access information or communicate with others at the same time.

Asynchronous Online Formats

Asynchronous formats include presentation of content through print documents, text messaging, images and animation, and audio and video media, alone or as part of multimedia modules. They facilitate interaction with learners through sent or posted text messages, and, more recently, voice messages. In general, asynchronous formats require minimal equipment and modest connection speeds and allow instructors and learners to access materials and exchange ideas at their own convenience.

Text and hypertext. Text passages have been and will continue to be the foundation of online instruction. Prepared in Hypertext Markup Language (HTML), text passages can elicit passive participation (read only) or active participation (linked to other information). Links, a hallmark of hypertext, can be used for elaboration (additional explanation), engagement (practice activities), comparison (simultaneous displays), or external sources (other Web sites) (Sanders, 2001).

Text messaging. The most common mechanism for online communication is text messaging. Text messaging is accomplished either through mail messages directly to one or more users or through bulletin boards that involve composing and posting a message in a location where it can be read by others (Kearsley, 2000). Mail systems usually include features, such as sending, replying, quoting, and attaching documents or images; they can be used for communication or for submission of assignments. Bulletin boards usually include features such as posting, replying, quoting, and attaching documents or images and also threading (organizing messages according to topics) and grouping (assigning individuals to specific groups); they can be used for making announcements or conducting discussions with large or small groups.

Documents. Print and image documents also are used in online instruction for presenting information or submitting assignments. Print documents are generally prepared as .doc files (primarily text) or .PDF files (text with images). Images are often prepared as .gif files (especially drawings or diagrams), but they also may be prepared as .jpg files, when higher resolution is needed (such as for photos). Large files, especially those that contained scanned text and images or photos with elaborate detail, may need to be optimized by removing unneeded information or compressed as a .zip file to avoid long download times.

Animations. Animations are sometimes used in online instruction to illustrate processes (such as a sequence of steps or events) or relationships among concepts (such as cause and effect). They may be prepared using a scripting language like Javascript as a .js file or in Macromedia Flash as a .swf file. Often they are designed as interactive activities that allow the learner to experience a new idea or model a process or practice some skill.

Streaming audio and video media. The advent of digital audio and video technology has enabled online instructors to present archived media content, either as downloadable files (wait to play on the desktop) or as streaming files (play in real time after a short buffer period). The real-time streaming protocol (RTSP) permits faster access but lower quality audio and video files (which are viewed as they are played), while the standard hypertext transfer protocol (HTTP) requires considerable time to download the files to the user's desktop but better quality images and sound (Bates & Poole, 2003). Downloadable files are best used for fairly short videos that have more detail, while streaming files are appropriate for interviews (audio only accompanied by a photo) or brief demonstrations (uncomplicated background and limited motion). When instruction involves archived video files of some length (such as full lectures), some authors recommend cataloging them on compact disks or digital videodisks rather than posting them on a Web site (Bates & Poole, 2003).

Using asynchronous online formats. Asynchronous formats can be used for self-study modules, courses, or workshops where individual learners work alone as well as for group activities, such as class discussions or team projects, that do not require real time exchanges among learners. Successful asynchronous instruction involves not only interactivity (with programs and content), but also interaction (with the instructor and other learners) (Palloff & Pratt, 2005). A hallmark of effective asynchronous instruction is the use of navigation aids that are easy to understand and use and are consistent throughout the course. In addition, to enhance compatibility and transferability in different learning environments and across computer platforms, content may be created using Extensible Markup Language (XML) and Shareable Content Object Reference Model (SCORM) standards (Moore & Kearsley, 2005).

Garrison and his colleagues (Garrison, Anderson, & Archer, 2003) have proposed a constructivist model of online instruction based on the creation of a community of learners supported through three mechanisms: (a) social presence, (b) cognitive presence, and (c) teaching presence. Social presence involves establishing a real and authentic identity through the use of first names, personal anecdotes, and humor. Cognitive presence involves constructing knowledge through the use of questioning strategies, prompts and exercises to engage learners in reflection, and explanation and discussion to integrate ideas and perspectives. Teaching presence involves structuring course activities to promote learning through organization, scheduling, management, and accommodation of individual needs. Moshinskie (2003) recommends both push strategies (require performance via monitoring) and pull strategies (inspire performance through incentives) to motivate participation in asynchronous learning activities.

SYNCHRONOUS ONLINE FORMATS

Synchronous online formats include presentation of content and/or interaction with learners through text chats, webcasting (live streaming in real time), audio conferencing, video conferencing, or virtual classrooms. Typically, synchronous formats require more sophisticated equipment and high bandwidth connectivity and require instructors and learners to be online at the same pre-scheduled times.

Text chats. Text chats are the simples and easiest to use synchronous interaction format, requiring no special equipment. Chat systems allow users to communicate by composing and posting text messages in real time that can be read and responded to by multiple users in real time (Kearsley, 2001). All chat participants can post a message at any point in time and messages appear on screen in the order in which they are submitted, accompanied by the name of the individuals who posted them. Most chat systems have more than one "room" so that multiple groups can interact simultaneously and some have functions that allow users to send private messages that cannot be seen by the group. Chats typically are logged to save a permanent text record that can be reviewed by the instructor or the learners throughout the course.

Webcasting. Streaming media technology also can be

used to create and transmit an audio and/or video stream of a live event (such as a class session or workshop presentation) in real time. In this process, known as webcasting (or web simulcasting if the event is also broadcast on television), the audio and video signal are digitized in real time and transmitted to a computer with server software that hosts the stream and allows access to multiple users (Sauer, 2001). Live transmission of instruction over the Internet using real-time streaming requires special equipment but is relatively inexpensive; however, interaction with learners during webcasts necessitates the use of other technology, such as a telephone conferencing system, which can raise the costs substantially (Bates & Poole, 2003).

Audioconferencing. Real time voice interactions are now possible online using a microphone and an audioconferencing program. Voice chats and discussion boards use Voice over Internet Protocol (VoIP) to transmit audio signal over the Internet alone or in combination with text tools, such as whiteboards or file sharing (Moore, 2004). Dedicated systems, such as Horizon Wimba (http://www.horizonwimba.com), iLinc Communications LearnLinc (http://ilinc.com), or Sonexis Conference Manager (http://www.sonexis.com/audio-conferencing.htm), allow conversations across multiple users and include moderator controls. Audioconferencing software permits access to a larger number of simultaneous sites, but requires users to speak in turn (half duplex) as opposed to the same time like a telephone which is full duplex) (Bates & Poole, 2003). Because the messages are digitized and transmitted as separate packages, audioconferencing can be interrupted by high traffic online or at peak use times.

Videoconferencing. The growing availability of high-speed connections and greater bandwidth is promoting greater interest in and use of videoconferencing applications that allow real time interactions with both image and voice. Videoconferencing requires each user to have a computer with video capability and a camera and may use point-to-point connection (between two users), multi-point connection (usually four to eight users), or a reflector site (similar to a server) where multiple images can be displayed and exchanged (Moore, 2004). Videoconferencing systems, such as Sorenson EnVision (http://www.sorenson.com), Polycom ClassStation (http://www.polycom.com), iVisit (http://www.ivist.com), Microsoft NetMeeting (http://www.microsoft.com/windows/netmeeting) and Apple iChat (http://www.apple.com/ichat) are software applications that can be used on any computer with reasonable processor speed and high bandwidth connection. At present, desktop video conferencing software delivers relatively poor image and sound quality and allows only a few sites to be connected simultaneously, but improvements in quality and distribution will increase the usefulness of this tool in years to come.

Virtual classrooms. The newest development in online instruction is the availability of synchronous interactive delivery systems termed virtual classrooms (McLaughlin, 2004). Programs, such as Macromedia Breeze (http://www.macromedia.com/software/breeze), IBM Lotus Virtual Classroom (http://lotus.com), or Webex Training Center (http://www.webex.com), create a virtual classroom environment that simulate face-to-face teaching and learning through real time interactions. Tools allow the instructor and learner to speak to the class, to use a shared whiteboard, to enter breakout rooms for discussion, and to poll the group and tally their responses. Virtual classrooms are the most expensive technology used for online instruction, costing upwards of $10,000 and often involving yearly licensing fees (Driscoll & Carliner, 2005).

Using synchronous online formats. Synchronous formats can be used for small group (two to six participants) activities, such as office hours, tutorials, and group discussions or meetings, or for large group (up to 100 participants) activities, such as classes or workshops and conferences (Boettcher, 2005). In general, successful synchronous instruction involves limiting group size or structuring activities to facilitate interactions (Driscoll & Carliner, 2005). A decision to use more expensive and complicated applications, such as webcasting or videoconferencing, should be based on available bandwidth, processing speed of user computers, and image quality produced by the compression scheme (Bates & Poole, 2003).

Instructors using synchronous Web-based instruction face two challenges (Hofmann, 2004): (a) understanding the tool well enough to know what can and cannot be done with it and (b) achieving balance between presentation and interaction. Hofmann recommends three strategies to enhance the effectiveness of synchronous delivery: (a) prework, (b) a leader guide, and (c) a participant guide. Prework is any activity that the instructor requires participants to complete to prepare for the live class session; this strategy allows the instructor to focus on illustration, explanation, and application and to address questions and issues. The leader guide is an outline or script that provides an organizational structure for the instructor to follow to stay on time and on topic during the live class session. The participant guide is a print document or set of presentation slides highlighting key points that the participant uses to follow the instructor's lesson.

WHY AND HOW IS ONLINE INSTRUCTION USED IN RURAL SPECIAL EDUCATION?

Online instruction is becoming an increasingly common mechanism for delivering preservice and inservice training in special education. Special educators were quick to understand the potential of technology-mediated distance education for personnel preparation, to address critical shortages (Howard, Ault, Knowlton, & Swall, 1992), to expand programs in low incidence disabilities (Spooner, 1996), and to enhance program access to individuals living and working in rural areas (Ludlow, 1995). Online instruction is used for distance delivery of teacher education programs to address teacher shortages, as well as professional development activities, to promote skill development and reduce teacher attrition.

Why is Online Instruction Used?

For all of its history, special education has been plagued by critical shortages of adequately trained and appropriately qualified personnel. These shortages have been due to an inadequate supply of new teachers produced by personnel preparation programs at institutions of higher education as well as by high rates of attrition for beginning special educators (Boe,

Cook, Bobbit, & Terhanian, 1998). Even the most recent data suggest that the supply-demand imbalance is likely to worsen significantly during the coming decade McLeskey, Tyler, & Flippin, 2004).

Rural schools in particular have faced many challenges in recruiting and retaining special education personnel. Early federal efforts to fund rural initiatives in special education are no longer active (Ludlow, 1998) and new programs addressing rural personnel needs are under-represented among current awardees (Rosenkoetter, Saceda, & Irwin, 2004). Graduates of personnel preparation programs are often reluctant to relocate or to return to rural areas (Hicks, 1994), and school administrators have reported difficulty recruiting personnel for positions in rural schools and retaining them in those positions once hired (National Association of State Directors of Special Education, 1996). Rural special educators also have higher rates of attrition, which may be due to stress and burnout (Westling & Whitten, 1996) as well as to less access to preservice preparation and professional development opportunities. (Storer & Crosswait, 1995). As a result of these combined factors, special educators have called for national policies to address the personnel needs of rural schools (Brownell, Rosenberg, Sindelar, & Smith, 2004). The use of online instruction for distance education programs is one mechanisms that has potential to reduce personnel shortages in rural areas.

How is Online Instruction Used?

Although distance education in teacher education in special education has a long history dating back over 20 years, the use of online instruction is a relatively new phenomenon. The initial reports of online instruction appeared in special education journals in the late 1990s but primarily described use of Web support for individual campus-based courses. A national study of distance education programs to prepare special education personnel reported in the professional literature from 1985 through 2000 (Ludlow & Brannan, 1999, dated that year but published in 2001) identified 32 programs, only one of which used online instruction. These researchers noted that most programs at that time were designed for low incidence areas and rural personnel, were primarily offered at the graduate level, and changed delivery systems as new technologies became available. A more recent national survey of distance delivery systems in low incidence disabilities only (Ludlow, Conner, & Schechter, 2005) located 28 programs, with 17 programs using Web support and 10 programs offered entirely online. This literature suggests a gradual progression from some support to total use and from single courses to full programs.

Web-enhanced courses. The literature contains early reports of online components developed to supplement face-to-face courses on campus or televised courses in distance education programs. Faculty at the University of Kansas used federal funds to develop the Online Academy, a series of modules in the three areas of behavior support, literacy instruction, and educational technology, made available online for use at other colleges and universities (Meyen, Lian, & Tangen, 1997). A Web-enhanced course in attention deficit disorders for special educators at the graduate level was offered at the

University of Florida using WebCT (Smith, Jordan, Corbett, & Dillon, 1999). A collaborative course involving multiple campuses in Wisconsin used electronic mail, discussion lists and boards, and Web Course in a Box to supplement distance delivery via compressed video (Hains, Conceicao-Runlee, Caro, & Marchel, 1999). Web support was used to enhance courses in an early intervention program delivered via telephone conferencing throughout rural Alaska (Ryan, 1999).

Web-based courses for preservice training. Individual courses conducted entirely online soon became possible with the availability of Web course management systems. The first fully Web-based course reported in the literature was delivered to doctoral students at the University of Kentucky using Top Class (Blackhurst, Hales, & Lahm, 1998). Online courses in disability characteristics and methods were offered at Jacksonville State University using Blackboard (Beard & Harper, 2002; Beard, Harper, & Riley, 2004). More recently, an introductory special education course was offered at East Carolina University using Blackboard (Steinweg, Davis, & Thurman, 2005) and a survey of behavior disorders course was offered using the Forum discussion board program and other online components at Indiana University (Knapczyk, Frey, & Wall-Marencik, 2005).

Online distance education programs. It was not long before special educators recognized the potential of online instruction as a primary distance delivery system for entire personnel preparation programs. The first program to utilize online instruction across an entire program was developed at the University of Utah, as part of a collaborative effort with other universities and school systems in the northwestern region of the country; it used desktop videoconferencing via CU-See Me to link these campus-based programs for live instruction (Spooner, Agran, Spooner, & Kiefer-O'Donnell, 2000). The University of Northern Colorado became the first institution to deliver a personnel preparation program entirely online; this graduate level program in vision impairments was delivered across a 14 state area (Ferrell, Persichitte, Lowell, & Roberts, 2001). Around the same time, the satellite-based distance education program at West Virginia University began to use real-time video streaming via RealMedia to allow students to access its courses in other areas of the country and in international locations without satellite access (Ludlow, 2003); in 2004, this program dropped the satellite component to rely exclusively on online instruction for all students.

Online professional development activities. The earliest use of online instruction for professional development of special education personnel was undertaken by the Council for Exceptional Children (CEC), which instituted a series of online seminars on current topic and issues in the field for continuing education credit (CEC, 1997). The Internet was used to facilitate the dissemination of resources, networking opportunities and inservice training for deaf educators (Johnson, 1997). Indiana University was among the first to explore the use of the Web to deliver inservice training in rural schools (Rodes, Knapczyk, Chapman, & Chung, 1999). In 2000, West Virginia University began offering four asynchronous summer courses for professional development of educators in rural areas of West Virginia and surrounding states using WebCT (Ludlow, Foshay, Brannan, Duff, & Dennison, (2002). Utah State University recently delivered an individual inservice course for teachers and paraeducators using videoconferencing with iVisit to reach rural areas of three states (Forbush & Morgan, 2004).

Online supervision and mentoring. The Internet also offers a mechanism for communication and interactions between college and university faculty and prospective or practicing teachers. Indiana University-Southeast pioneered the development of online mentoring of beginning special educators in an induction program (Shea & Babione, 2001). Utah State University was the first to harness the power of desktop conferencing to maintain contact with student teachers and their supervisors at field sites in remote areas, first through Sorenson EnVision and later through the Polycom system (Binner/Falconer & Lignugaris/Kraft, 2002). The University of North Texas also has utilized desktop videoconferencing using EnVision with multi-point access for supervision of field-based practicum experiences (Pemberton, Cereijo, Tyler-Wood, & Rademacher, 2003).

Concerns about Online Instruction in Special Education

The use of online instruction for personnel preparation in special education has raised important questions yet to be adequately addressed about its application in this field. Spooner and his colleagues (Spooner, Spooner, Algozzine, & Jordan (1998) expressed concern that asynchronous online instruction may lack some of the interpersonal dimensions of the face-to-face classroom that could be essential to the development of teachers. Kiefer-O'Donnell and Spooner (2002) lamented the rapid implementation of distance education programs without an adequate knowledge base to support best practices. Meyen and his colleagues (2002), addressing online instruction specifically, called for improved research. Johnson (2004) warned that, when they simply translate their face-to-face course components into the online environment, instructors may produce what is little more than an electronic correspondence course and fail to capitalize on the advantages of online learning. Maddux (2004) identified and dispelled 10 myths about the use of online instruction; he indicated that online programs are growing at a rapid pace and institutions of higher education must offer such programs or lose enrollments to private providers. Finally, in her review of the literature on the effectiveness and impact of technology-mediated distance education (including online instruction) in special education, Ludlow (2005) noted the lack of sound design and the absence of conclusive findings and called for additional research to clarify the effectiveness of specific technologies in producing learning outcomes and to document the impact of distance delivery models on special education teacher shortages.

WHAT ARE ISSUES IN USING ONLINE INSTRUCTION?

Distance education, in general, and online instruction, in particular, have generate considerable controversy among both academics and the public at large. Van Dusen (2000) conceptualized distance education (including online instruction) as presenting both opportunities and challenges with respect to three factors: (a) access, (b) quality, and (c) cost. Some authors have criticized online instruction for sacrificing quality in the interests of access and cost (Bates, 2000). Others, however, have asserted that online instruction can enhance both access and quality even while reducing costs (Weigel, 2002). Both proponents and opponents agree that many issues must be addressed if the potential of online instruction for increasing access, maintaining quality, and controlling costs is to be realized.

Issues Related to Access

One of the major goals of using online instruction for distance education in special education is to increase access of personnel preparation to rural areas. Consequently, a primary concern is whether online instruction enhances or impedes access to training programs for individuals or groups. Three issues related to access are (a) the availability of technology to individuals and communities, (b) the familiarity of learners with new technologies, and (c) the accessibility of online formats for people with disabilities.

Technology availability. Some educators have expressed concern that using online instruction in general or specific online instructional formats (especially synchronous formats such as videoconferencing) may not in fact enhance access to all areas and people. Studies of Internet use have suggested the existence of a "digital divide" related to geographic area, income status, and personal characteristics (Van Dusen, 2000). Several federal policy initiatives have been implemented to address regional inequities and improve access to the Internet and computers in low-income areas and rural locales (Moore & Kearsley, 2005).

Until fairly recently, Internet service was not available in many rural communities, and, even now in some locations, service is minimal or easily disrupted. Available data, however, suggest that the technical problems experienced by rural special educators during online instruction have been largely due to bandwidth issues, which causes difficulty in downloading large files (Menlove & Lignugaris/Kraft, 2004) or viewing media (Ludlow, Foshay, Brannan, Duff, & Dennison, 2002). Some authors have suggested that online instruction be designed to minimize the need for specialized technology or technical skills so students can access the content more readily (Meyer, Tangen, & Lian, 1999).

Learner familiarity. New technologies only can be successful if users can are skilled and comfortable enough to make effective use of them for teaching and learning. Some authors have argued that totally online courses and programs may not meet the needs of nontraditional students, especially women and members of culturally diverse groups who may prefer some face-to-face instruction (Carriuolo, 2002). Others have noted that learners with a history of traditional, passive educational experiences (lectures, print materials) tend to be uncomfortable and even resistant to newer, more active methods of instruction, such as threaded discussions or virtual classrooms (McCormack & Jones, 1998). Palloff and Pratt (2003) suggest that younger or inexperienced online learners benefit from higher structure, concrete learning experiences, and required sequential tasks, while older, more experienced learners are more comfortable with independent study, open-ended tasks, and self-designed learning.

Preservice and especially inservice special educators do not always have the level of technology skills needed to engage successfully in online learning activities. Some have recommended that explicit instruction in technology skills be embedded in online courses (Schnorr, 1999). Others have suggested that learners need both online orientation with practice exercises and on going technical support services to develop skills for online learning (Collins, Schuster, Ludlow & Duff, 2002). Still others have required students to attend distance learning workshops or seminars prior to enrollment in an online program (Menlove & Lignugaris/Kraft, 2004).

Accessibility for people with disabilities. The provisions of the Americans with Disabilities Act (ADA) of 1990 apply to distance education, so colleges and universities must insure that learners with disabilities have access to such programs and that reasonable accommodations are made to meet their individual needs (see http://www.section 508.gov). Full participation by individuals with disabilities in distance education courses and programs requires accessible instructional materials, use of assistive hardware and software, and availability of technical support (Palloff & Pratt, 2003). Accessible materials include standards-compliant Web pages; use of text alternatives, such as transcripts and descriptions of audio and video media; and additional participation options if some formats (for example, text chats or audio conferences) cannot be accessed. Developers of online courses and programs can consult the World Wide Web Consortium's guidelines for web content accessibility (World Wide Web Consortium, 1999) for information about how to make content and navigation accessible to individuals with disabilities. These guidelines address the use of style sheets to format content pages, alt tags to access text (print descriptions of images) or non-text (spoken description of images) equivalents to multimedia content, and alternative access to dynamic content (java scripts and applets, interactive media).

Issues Related to Quality

Another major goal in the use of online instruction for distance education is to ensure that such courses and programs achieve the desired quality of learning experiences. As a result, concern has been expressed over how well online instruction measures up to standards for higher education courses and programs. Issues related to quality include (a) granting accreditation of online programs, (b) evaluating the effectiveness of online instruction, and (c) maintaining standards of academic integrity.

Program accreditation. Some of the standards developed to ensure the quality of institutions of higher education through accreditation bodies may not be easily applied to distance education courses and programs. Educators have expressed concern that online instruction threatens core academic values (Easton, 2000), relinquishing faculty control to instructional designers and technical support personnel and diminishing opportunities for meaningful and interactive learning experiences. To address this problem, the Institute for Higher Education Policy (2001) has published a list of

24 quality indicators for Internet-based distance education in seven categories of institutional support, course development, teaching/learning, course structure, student support, faculty support, and evaluation and assessment. The Council of Regional Accrediting Commissions (2000) now also offers a set of adapted standards to evaluate the quality of courses and programs delivered via online instruction for use in accreditation reviews.

Effective instruction. A great deal of time and effort has gone into collecting data to provide whether distance education, including online delivery models, is as effective as face-to-face instruction. The failure of such research to find any major differences in either learning outcomes or learner satisfaction has been termed the "no significant difference phenomenon" (Russell, 1999). Meyer (2002) summarized available reviews of the distance education research and confirmed that outcomes are about the same no matter what technology is used for distance education. She argued that concerns about quality are really about discomfort with change and reflect efforts to preserve the status quo rather than to ensure effectiveness of online courses and programs. Arbaugh and Hiltz (2005) have recommended that future research into the effectiveness of online instruction needs to be designed for greater control over variables as well as to investigate how specific online formats may be related to various learning outcomes.

There have been several reports in the professional literature in special education that evaluated individual online courses in personnel preparation or for professional development. However, there have been no studies reporting the results of distance education programs delivered entirely online at this point in time. A recent review of the research on distance education in special education, including online instruction, concluded that the no significant difference phenomenon occurs in this field as well (Ludlow, 2005). Therefore, at present there is only minimal (and somewhat inconclusive) evidence of the effectiveness of online instruction in preparing special education personnel.

Academic integrity. A common objection to online instruction on the part of faculty and institutions is that it provides more opportunities for learners to breach standards of academic integrity. However, there is no evidence that cheating and plagiarism occur more frequently in online learning; in fact, some studies suggest that it occurs at about the same rate as in face-to-face courses (Kellogg, 2002). Palloff and Pratt (2003) recommend that instructors take measures to prevent and detect academic dishonesty, including designing assignments that are original and open-ended and cannot be copies from other materials. In future, course management systems may be set up to use biometric devices (such as eye scans or voice prints) to identify authorized users (Dabbaugh & Bannan-Ritland, 2005).

Issues Related to Cost

A final goal for the use of online instruction for distance education is to control the cost of personnel preparation and professional development programs, both resource costs and personnel costs. A major concern is whether online instruction helps institutions to manage costs or simply transfers costs from physical to technical infrastructures. Three issues related to the cost of online instruction are (a) personnel effort needed to develop and deliver online instruction, (b) resource costs associated with development and delivery of instruction, and (c) debates over the ownership of intellectual property.

Personnel effort. Online instruction requires considerably more time and effort to develop, which has significant implications for faculty workload and compensation. A variety of commentators have noted that online courses require more time for both development and delivery of instruction (Brown, 1998; Killian, 1997). Development time depends on the complexity of the course but usually involves a significant amount of advance preparation (Meyer, 2002). When development is done by the individual instructor, the institution may offer incentives, such as release time, stipends, or salary increments. When development requires a team effort, the program may need to hire personnel, such as media producers and computer programmers to balance workloads. Delivery time is increased largely by the fact that students expect (and instructors feel compelled to comply) prompt, detailed, and individualized feedback in the 24/7 online environment (Boettcher, 2004). Estimates of the appropriate number of students per course range from 15 to 30, with the average around 20 (Kearsley, 2000). Courses with higher enrollments typically rely on some form of grading assistance to enable the instructor to manage the added work. At some institutions, faculty receive extra pay or other benefits to compensate them

Resource costs. Online instruction can consume a significant amount of institutional resources for both development or delivery. Courses can be inexpensive or quite expensive to develop, with estimates in the range of $50,000 to $100,000 per course for team development (Downes, 1998). Greater use of multimedia generally increases both development and delivery costs (Bates & Poole, 2003). However, new open content initiatives that facilitate sharing of content across faculty and institutions may help to control costs through leasing agreements (Boettcher, 2004). In general. the costs of online instruction will vary with factors such as number of courses, frequency of revision, type of technology used, support services, degree of interactivity, enrollment, and completion and attrition rates (Jung, 2003). Although online instruction has the potential to produce economies of scale, achieving cost control may only be possible through collaborative agreements across colleges and universities (Van Dusen, 2000).

Intellectual property ownership. The significant amount of instructor time and institutional resources needed for online instruction raises new questions about ownership of intellectual property in higher education settings. Intellectual property rights define the boundary between the faculty member and the institution and between academia and the marketplace (McSherry, 2003). On the one hand, academic traditions of ownership suggest that instructional materials belong to the faculty member who created them. On the other hand, institutions regard online courses as work for hire and a return on investment in personnel salaries and equipment purchases. Institutions should develop specific policies related to intellectual property rights that allocate ownership through contracts or royalty agreements and designate who controls repeated use (Primo & Lesage, 2001). Programs and instructors also need to take steps to prevent unauthorized use or outright theft of their work by using copy protection mechanisms (Ko & Rossen, 2004).

EMERGING TRENDS IN ONLINE INSTRUCTION

The use of online instruction is so new and so different in many respects from academic traditions that its impact on higher education is open to speculation. Three emerging trends will influence future use of online instruction: (a) the threat of competition among institutions for new audiences and markets to expand enrollment or increase revenue, (b) the opportunities for program outreach to international areas, and (c) the design of ever smaller devices for truly mobile computing.

COMPETITION AMONG INSTITUTIONS

Online courses and programs know no bounds of time or space, so they offer colleges and universities the opportunity to reach out to audiences beyond their traditional service areas. These outreach efforts raise the threat of competition for learners between institutions and with non-academic entities (Katz, & Associates, 1999). Some commentators have suggested that the rapid growth of private, for-profit providers of totally online programs (such as the University of Phoenix and Walden University) signals a death knell for traditional colleges and universities (Mason, 2003). However, there is some evidence that students prefer online instruction that is presented by traditional institutions of higher education rather than private, for-profit companies (Lorenzo, 2000). To reduce competition and foster collaboration in online instruction, some institutions have entered into consortium agreements (such as the Western Governor's University, the Southern Regional Education Board's Electronic Campus, or the Kentucky Virtual Campus) (Moore & Kearsley, 2005).

INTERNATIONAL OUTREACH

The widespread availability of Internet access offers the opportunity for institutions of higher education to offer courses and programs with an international scope. New possibilities for digitization of materials and events and instantaneous worldwide transmission will foster the globalization of education in the years to come (Weigel, 2002). However, international distance education will present new challenges for colleges and universities, such as the need for worldwide program accreditation (Jarvis, 2000). In addition, extension of program offerings to international locations raises issues related to the language and culture of instruction (Mason, 2003). Language barriers may be solved in the future by emerging options for automated language translation (Kearsley, 2000). Cultural barriers must be addressed by instructional design that both accommodates and reflects diversity (Goodfellow, Lea, Gonzalez, & Mason, 2001).

MOBILE COMPUTING

The increasing availability of wireless communication through cellular phones, personal digital assistants (PDAs), and handheld computers is opening up new possibilities for mobile computing, including online instruction. Mobile learning or m-learning (Driscoll & Carliner, 2005) involves wireless connectivity, portable devices, and a self-contained power supply, while desktop learning or e-learning involves fixed lines, full screen computers, and an external source of power. M-learning differs from e-learning in the amount and type of content and the degree of interactivity due to the differences in bandwidth, display sizes, and input mechanisms. However, m-learning also requires that content be mobile as well, which can be accomplished through the use of XML files or "podcasting" (Villano, 2005). The possibilities inherent in mobile computing will give new meaning to "any time, any place" learning opportunities.

CONCLUSION

It seems clear that online instruction is not only an idea whose time has come, but it is an innovation that is here to stay. The Sloan Consortium's second annual study of online instruction in higher education (Allen & Seaman, 2004) revealed that enrollments in online courses and programs have continued to grow exponentially and to outstrip increases in general student enrollment in the nation's colleges and universities. The study also showed that institutions report the quality of instruction is equal to or better than face-to-face teaching and that students are as or more satisfied with online learning as with traditional instruction, especially in graduate and professional programs and at larger schools. Boettcher (2004) predicts that, with the coming decade, all courses at all institutions of higher education will blend face-to-face and online instruction to incorporate the most effective features of each delivery system. Thoughtful applications of online instruction can help institutions of higher education expand their access and increase their revenues while maintaining program quality, assisting them to achieve a balance between being "market smart and mission centered" (Zemsky, Wegner, & Massy, 2005). These developments will blur the distinction between campus-based and distance education programs and position more programs to extend their reach beyond traditional boundaries. No doubt, increasing use of online instruction for personnel preparation in special education holds great promise for increasing access to quality preservice and inservice programs to special educators working and living in rural areas.

REFERENCES

Allen, I. E., & Seaman, J. (2004). *Entering the mainstream: The quality and extent of online education in the United States, 2003 and 2004*. Needham, MA: The Sloan Consortium. Retrieved December 1, 2004 from http://www. sloan-c.org.

Arbaugh, J. B., & Hiltz, S. R. (2005). Improving quantitative research on ALN effectiveness. In S. R. Hiltz & R. Goldman (Eds.), *Learning together online: Research on asynchronous learning networks* (pp. 81-102). Mahwah, NJ: Lawrence Erlbaum Associates.

Bates, A. W. (2000). *Managing technological change: Strategies for college and university leaders*. San Francisco: Jossey-Bass.

Bates, A. W., & Poole, G. (2003). *Effective teaching with technology in higher education: Foundations for success*. San Francisco: Jossey-Bass.

Beard, L. A., & Harper, C. (2002). Student perceptions of online versus campus instruction. *Education, 122*(4), 658-665.

Beard, L. A., Harper, C., & Riley, G. (2004). Online versus on-campus instruction: Student attitudes and perceptions. *TechTrends, 48*(6), 29-31.

Bernard, R. M., Abrami, P. C., Lou, Y., Borokhovski, E., Wade, A., Wozney, L., Wallet, P. A., Fiset, M., & Huang, B. (2004). *Review of Educational Research, 74*(3), 379-439.

Binner/Falconer, K., & Lignugaris/Kraft, B. (2002). A qualitative analysis of the benefits and limitations of using two-way conferencing technology to supervise preservice teachers in remote locations. *Teacher Education and Special Education, 25*(4), 368-384.

Blackhurst, A. E., Hales, R. M., & Lahm, E. A. (1998). Using an education server software system to deliver special education coursework via the Worldwide Web. *Journal of Special Education Technology, 13*(4), 78-98.

Boe, E. E., Cook, L. H., Bobbitt, S. A., & Terhanian, G. (1998). The shortage of fully certified teachers in special and general education. *Teacher Education and Special Education, 21*(1), 1-21.

Boettcher, J. V. (2005). Designing for the virtual interactive classroom. *Campus Technology, 18*(9), 20-23.

Boettcher, J. V. (2004). Distance education: Are we there yet? *Campus Technology, 18*(3), 22-25, 26.

Brown, B. M. (1998). Digital classrooms: Some myths about developing new educational programs using the Internet. *T. H. E. Journal, 26*(5), 56-59.

Brownell, M. T., Rosenberg, M. S., Sindelar, P. T., & Smith, D. D. (2004). Teacher education: Toward a qualified teacher for every classroom. In Sorrells, A. M., Rieth, H. J., & Sindelar, P. T. (2004). *Critical issues in special education: Access, diversity, and accountability, and diversity* (pp. 243-257). Boston: Allyn and Bacon.

Calhoun, T. (2005). DRM: The challenge of the decade. *Campus Technology, 18*(7), 18-19.

Carriuolo, N. (2002). The nontraditional undergraduate and distance learning: Is higher education providing a portal or just a keyhole to social and economic mobility? *Change, 34*(6), 56-62.

Collins, B. C., Schuster, J. W., Ludlow, B. L., & Duff, M. (2002). Planning and delivery of online coursework in special education. Teacher Education and Special Education, 25(2), 171-186.

Council for Exceptional Children. (CEC). (1997). CEC introduces electronic study group. *Teaching Exceptional Children, 29*(6), 11.

Council of Regional Accrediting Commissions. (2000). *Statement of the regional accrediting commissions on the evaluation of electronically offered degree and certificate programs and guidelines for the evaluation of electronically offered degree and certificate programs.* Retrieved January 10, 2005 from http://www.wiche.edu/edu/wcet/resources/publications/guidelines.pdf.

Dabbagh, N., & Bannan-Ritland, B. (2005). *Online learning: Concepts, strategies, and applications.* Upper Saddle River, NJ: Pearson Education.

Day, J. N., & Sebastian, J. P. (2002). Preparing vision specialists at a distance: A qualitative study on computer-enhanced learning. *Journal of Visual Impairment and Blindness, 96*(11), 796-808.

Downes, S. (1998). The future of online learning. *Online Journal of Distance Learning Administration, 1*(3). Retrieved December 1, 2004 from http://www.westga.edu/`distance/downes13.html.

Driscoll, M., & Carliner, S. (2005). *Advanced Web-based training strategies: Unlocking instructionally sound online learning.* San Francisco: Pfeiffer.

Easton, J. S. (2000). *Core academic values, quality, and regional accreditation: The challenge of distance learning.* Washington, DC: Council for Higher Education Accreditation.

Ferrell, K. A., Persichitte, K. A., Lowell, N., & Roberts, S. (2001). The evolution of a distance delivery system that supports content, students, and pedagogy. *Journal of Vision Impairments and Blindness, 95,* 597-609.

Forbush, D. E., & Morgan, R. L. (2004). Instructional team

training: Delivering live, Internet courses to teachers and paraprofessionals in Utah, Idaho, and Pennsylvania. *Rural Special Education Quarterly, 23*(2), 9-17.

Garrison, D. R., Anderson, T., & Archer, W. (2003). A theory of critical inquiry in online distance education. In M. G. Moore & W. G. Anderson (Eds.), *Handbook of distance education* (pp. 113-127). Mahwah, NJ: Lawrence Erlbaum Associates.

Goodfellow, R., Lea, M., Gonzalez, F., & Mason, R. (2001). Opportunity and e-quality: Intellectual and linguistic issues in global distance education. *Distance Education, 22*(1), 65-84.

Griffin-Shirley, N., Almon, P., & Kelley, P. (2002). Visually impaired personnel preparation program: A collaborative distance education model. *Journal of Visual Impairment and Blindness, 96*(4), 233-245.

Hains, A. H., Conceicao-Runlee, S., Caro, P., & Marchel, M. A. (1999). Collaborative course development in early childhood special education through distance learning. *Early Childhood Research & Practice, 1*(1). Retrieved October 15, 2004 from http://ecrp.uiuc.edu/v1n1/hains.html.

Hicks, J. (1994). *Special education in rural areas: Validation of critical issues by selected state directors of special education: Final report.* Alexandria: National Association of State Directors of Special Education (ERIC Document No. ED 738 767).

Hofmann, J. (2004). *Live and online! Tips, techniques, and ready-to-use activities for the virtual classroom.* San Francisco: Pfeiffer.

Howard, S. W., Ault, M. M., Knowlton, H. E., & Swall, R. A. (1992). Distance education: Promises and cautions for special education. *Teacher Education and Special Education, 15*(4), 275-283.

Institute for Higher Education Policy. (2001). *Quality on the line: Benchmarks for success in Internet-based distance education.* Washington, DC: National Education Association. Retrieved December 1, 2004 from http://www.nea.org/nr/nr000321.html.

Jarvis, P. (2000). The changing university: Meeting a need and needing to change. *Higher Education Quarterly, 54*(1), 43-67.

Johnson, H. A. (1997). Internet solutions for isolation: Educational resources and professional development opportunities for educators of deaf and hard of hearing students. *Rural Special Education Quarterly, 16*(2), 33-41.

Johnson, L. R. (2004). Research-based online course development for special education teacher preparation. *Teacher Education and Special Education, 27*(3), 207-223.

Jordan, L., Smith, S., Dillon, A. S., Algozzine, B., Beattie, J., Spooner, F., & Fisher, A. L. (2004). Improving content and technology skills in ADD/ADHD via a Web-enhanced course. *Teacher Education and Special Education, 27*(3), 231-239.

Jung, I. (2003). Cost-effectiveness of online education. In M. G. Moore & W. G. Anderson (Eds.), *Handbook of distance education* (pp. 717-726). Mahwah, NJ: Lawrence Erlbaum Associates.

Katz, R. N., & Associates. (1999). *Dancing with the devil: Information technology and the new competition in higher education.* San Francisco: Jossey-Bass.

Kearsley, G. (2000). *Online education: Learning and teaching in cyberspace.* Belmont, CA: Wadsworth.

Kellogg, A. P. (2002, February 1). Students plagiarize less than many think, a new study finds. Chronicle of Higher Education, A2-A3.

Kilian, C. (1997). F2F – Why teach online? *Educom Review, 32*(4), 31-35.

Knapczyk, D. R., Frey, T. J., & Wall-Marencik, W. (2005). An evaluation of Web conferencing in online teacher preparation. *Teacher Education and Special Education, 28*(2), 114-124.

Ko, S., & Rossen, S. (2004). *Teaching online: A practical guide* (2nd ed.). Boston: Houghton Mifflin.

Lifter, K., Kruger, L., Okun, B., Tabol, C., Poklop, L., & Shishmanian, E. (2005). Transformation to a Web-based preservice training program: A case study. *Topics in Early Childhood Special*

Education, 25(1), 15-24.

Lock, R. H. (2001). Using Web-based information to facilitate inclusion practices in rural communities. *Rural Special Education Quarterly, 20*(4), 7-19.

Lorenzo, G. (1998). *Study shows big demand for online education*. Retrieved December 1, 2004 from http://www.edpath.com/study.htm.

Ludlow, B. L. (2003). An international outreach model for preparing early intervention and early childhood special education personnel. *Infants and Young Children, 16*(3), 238-248.

Ludlow, B. L. (1995). Distance education applications in rural special education: Where we've been and where we're going. *Rural Special Education Quarterly, 14*(2), 45-52.

Ludlow, B. L. (1998). Preparing special education personnel for rural schools: Current practices and future directions. *Journal of Research in Rural Education, 14*(2), 57-75.

Ludlow, B. L. (2005). Technology-mediated distance education: Current practices and future trends. In D. L. Edyburn, K. Higgins, & R. Boone (Eds.), *Handbook for special education technology research and practice* (pp. 1-25). Whitefish Bay, WI: Knowledge by Design.

Ludlow, B. L., & Brannan, S. (1999, published in 2001). Distance education programs for preparing personnel for rural areas: Current practices, emerging trends, and future directions. *Rural Special Education Quarterly, 18*(3/4), 5-20.

Ludlow, B. L., Foshay, J. D., Brannan, S. A., Duff, M. C., & Dennison, K. E. (2002). Updating knowledge and skills of practitioners in rural areas: A Web-based model. *Rural Special Education Quarterly, 21*(1), 33-43.

Maddux, C. D. (2004). Developing online courses: Ten myths. *Rural Special Education Quarterly, 23*(2), 27-32.

Mason., R. (2003). Global education: Out of the ivory tower. In M. G. Moore & W. G. Anderson (Eds.), *Handbook of distance education* (pp. 743-752). Mahwah, NJ: Lawrence Erlbaum Associates.

McCormack, C., & Jones, D. (1998). *Building a Web-based communication system*. New York: John Wiley & Sons.

McSherry, C. (2003). *Who owns academic work? Battling for control of intellectual property*. Cambridge, MA: Harvard University Press.

McLaughlin, M. K. (2004). Online learning. *PC Magazine, 23*(16), 124.

McLeskey, J., Tyler, N. C., & Flippin, S. S. (2004). The supply and demand of special education teachers: A review of research regarding the chronic shortage of special education teachers. The *Journal of Special Education, 38*(1), 5-21.

Menlove, R., & Lignugaris/Kraft, B. (2004). Preparing rural distance education preservice educators to succeed. *Rural Special Education Quarterly, 23*(2), 18-26.

Meyen, E. L., Lian, C. H. T., & Tangen, P. (1997). Teaching online courses. *Focus on Autism and Other Developmental Disabilities, 12*(3), 166-174.

Meyen, E. L., Tangen, P., & Lian, C. H. T. (1999). Developing online instruction: Partnership between instructors and technical developers. *Journal of Special Education Technology, 14*(1), 18-31.

Meyen, E. L., & Yang, C. H. (2005). Online staff development for teachers: Multi-state planning for implementation. *Journal of Special Education Technology, 20*(1), 41-54.

Meyer, K. A. (2002). *Quality in distance education: Focus on online learning*. San Francisco: Jossey-Bass.

Miller, M. D., Brownell, M. T., & Smith, S.W. (1999). Factors that predict teachers staying in, leaving, or transferring from the special education classroom. *Exceptional Children, 65*, 201-218.

Moore, C. (2004). Web conferencing meets desktops. *InfoWorld, 26*(18), 20.

Moore, M., & Kearsley, G. (2005). *Distance education: A systems view* (2nd. ed.). Belmont, CA: Thomson Wadsworth.

Morgan, R. L., Forbush, D. E., & Nelson, J. (2004). Live, interactive paraprofessional training using Internet technology: Description and evaluation. *Journal of Special Education Technology, 19*(3), 25-33.

Moshinskie, J. (2003). Organizational best practices for preparing e-learners. In G. M. Piskurich (Ed.), *Preparing learners for e-learning* (pp. 91-100). San Francisco: Jossey-Bass/Pfeiffer.

National Association of State Directors of Special Education. (1996, December). Special educator supply and demand in rural areas: Facing the issues. *Liaison Bulletin, 26*, 1-7.

Palloff, R. M., & Pratt, K. (2005). *Collaborating online: Learning together in the community*. San Francisco: Jossey-Bass.

Palloff, R. M., & Pratt, K. (2001). *Lessons from the cyberspace classroom: The realities of online teaching*. San Francisco: Jossey-Bass.

Palloff, R. M., & Pratt, K. (2003). *The virtual student: A profile and guide to working with online learners*. San Francisco: Jossey-Bass.

Pemberton, J. B., Cereijo, M. V. P., Tyler-Wood, T., & Rademacher, J. (2004). Desktop videoconferencing: Examples of applications to support teacher training in rural areas. *Rural Special Education Quarterly, 23*(2), 3-8.

Perrin, D. G., & Perrin, E. (2003). New copyright law for distance education. *USDLA Journal, 17*(1). Retrieved December 1, 2004 from http://www.usdla.org/html/journal/JAN03_Issue/editor.html.

Pindiprolu, S. S., Peterson, S. M., Rule, DS., & Lignugaris-Kraft, B. (2003). Using Web-mediated experiential case-based instruction to teach functional behavioral assessment skills. *Teacher Education and Special Education, 26*(1), 1-16.

Primo, L. H., & Lesage, T. (2001). Survey of intellectual property issues for distance learning and online educators. *USDLA Journal, 15*(2). Retrieved December 1, 2004 from http://www.usdla.org/html/journal/FEB01_Issue/article03.html.

Rodes, P., Knapczyk, D., Chapman, C., & Chung, H. (1999). Involving teachers in Web-based professional development. *T.H.E. Journal, 27*(10), 94-102.

Russell, T. L. (1999). *The no significant difference phenomenon.* Raleigh, NC: North Carolina State University, Office of Educational Telecommunications.

Ryan, S. (1999). Alaska's rural early intervention preservice training program. *Rural Special Education Quarterly, 18*(3/4), 21-28.

Sanders, W. B. (2001). *Creating learner-centered courses for the World Wide Web.* Boston: Allyn and Bacon.

Sauer, J. (2001). The stream team: A digital video format primer. *Emedia, 14*(10), 36-41.

Schnorr, J. M. (1999). Developing and using technology for course delivery. *Teacher Education and Special Education, 22*(2), 114-122.

Shea, C., & Babione, C. (2001). *The Electronic Enhancement of Supervision Project (EESP).* ERIC Document Reproduction Services No. ED 463 126.

Smith, S. B., Jordan, K., Corbett, N. L., & Dillon, A. S. (1000). Teachers learn about ADHD on the Web: An online graduate special education course. *Teaching Exceptional Children, 31*(6), 20-27.

Spooner, F. (1996). Personnel preparation: Where we have been and where we may be going in severe disabilities. *Teacher Education and Special Education, 19*(3), 213-215.

Spooner, F., Agran, M., Spooner, M., & Kiefer-O'Donnell, R. (2000). Preparing personnel with expertise in severe disabilities in the electronic age: Innovative programs and technologies. *Journal of the Association for Persons with Severe Handicaps, 25*(2), 92-103.

Spooner, F., Spooner, M., Algozzine, B., & Jordan, L. (1998). Distance education and special education: Promises, practices, and potential pitfalls. *Teacher Education and Special Education, 21*(2), 121-131.

Steinweg, S. B., Davis, M. L., & Thomson, W. S. (2005). A comparison of traditional and online instruction in an Introduction to Special Education Course. *Teacher Education and Special Education, 28*(1), 62-73.

Storer, J. H., & Crosswait, D. J. (1995). Delivering staff development to the small rural school. *Rural Special Education Quarterly, 14*(3), 23-30.

Van Dusen, G. C. (2000). *Digital dilemma: Issues of access, cost, and quality in media-enhanced distance education*. San Francisco: Jossey-Bass.

Villano, M. (2005). Imagination on the move. *Campus Technology, 18*(11), 32-37.

Weigel, V. B. (2002). *Deep learning for a digital age: Technology's untapped potential to enrich higher education*. San Francisco: Jossey-Bass.

Westling, D. L., & Whitten, T. M. (1996). Rural special education teachers' plans to continue or leave their teaching positions. *Exceptional Children, 62*(4), 319-335.

World Wide Web Consortium. (1999). *Web content accessibility guidelines version 1.0*. Retrieved December 1, 2004 from http://www.w3.org/TR/WAI-WEBCONTENT/#toc.

Zemsky, R., Wegner, G. R., & Massy, W. F. (2005, July 15). Today's colleges must be market smart and mission centered. *The Chronicle of Higher Education*, B6-B7.

Chapter 2

PLANNING AND DELIVERING ONLINE INSTRUCTION

BELVA C. COLLINS
University of Kentucky
CATHY GALYON-KERAMIDAS
West Virginia University

Online instruction is growing in popularity as a way for institutions of higher education to deliver coursework to students who are geographically remote from campuses or who need flexibility in order to schedule coursework around work and family obligations. An online delivery format has been especially advantageous in preparing special education personnel, both at the preservice and inservice levels, who are teaching in remote sites and need coursework to obtain licensure or to update skills. Some institutions that have been involved in distance education over time have either replaced other modes of distance delivery with online instruction or have added online instruction as a supplement to other types of technology. For example, West Virginia University's program in severe/multiple disabilities and early intervention/

early childhood special education has transitioned over time from using satellite delivery to interactive video delivery to online delivery (Ludlow, Foshay, Brannan, Duff, & Dennison, 2002), and the Collaborative Teacher Education Program in Indiana has used a hybrid audiographic system consisting of telephone conferencing with computer graphics to provide special education coursework for rural students (Knapczyk, Rodes, Haejin, & Chapman, 1999).

The advantages of online delivery are numerous (e.g., Beard & Harper, 2002; Ludlow et al., 2002):

1. Coursework can be offered to individuals or groups across a broad geographic region and in remote sites.
2. Students and instructors do not have to spend time or funds on travel.
3. A professional community can be formed to provide mutual support in the field.
4. The technology skills of students can be developed and enhanced over time.
5. Enrollment does not have to be limited based on seating capacity or availability of classroom sites.
6. Shy students may be less inhibited about participating in class interactions.
7. Class participation is not limited to scheduled course hours or sites.
8. The disabilities of students are not disclosed unless students desire to make them known, resulting in less stigma

and a more level field of participation.

In spite of these advantages, online instruction also poses numerous challenges (e.g., Bruce & Hwang, 2001; Ludlow et al., 2002; Meyen & Lian, 1997b):

1. Investment of personnel and finances must be made prior to course delivery.
2. Development and instruction are both time-consuming and labor intensive in an academic environment that values traditional modes of delivery.
3. The technical equipment and capabilities of students may be inadequate for advanced formats.
4. Students may require orientation to new forms of technology and ongoing technical support throughout coursework.
5. Technical problems, such as server outages, high server traffic, and incompatible equipment, may occur.
6. Costs of equipment and necessary updates may be expensive.
7. Accessibility for students with various types of disabilities must be addressed.
8. The learning styles of some students may be better suited to face-to-face interactions.
9. Instruction may be less personal and offer less socialization.
10. Some students, especially non-traditional students, may

be reluctant to attempt coursework using technology.

11. Student privacy issues (e.g., grades) may be violated.

12. Instructors may tend to focus on the technology instead of the content, failing to consider the type of technology that is best suited to the content.

13. Course content is more open to public scrutiny since it can be accessed over time.

14. Course revisions are time-consuming and difficult to make once the course begins.

15. Problems may arise when students enroll under the misconceptions that online courses are less rigorous and that it is easier to engage in plagiarism and cheating in the absence of face-to-face contact.

16. Universities may push high course enrollments as a cost-saving measure in spite of what it is feasible for instructors to manage.

Beard and Harper (2002) gave graduate students in special education at Jacksonville State University the opportunity to participate in a traditional class or to access course notes and supplemental materials posted online within a BlackBoard course management system. Students had the opportunity to experience both formats since the online component was not available until mid-term. When the researchers compared the attitudes and opinions of students toward the in-class and online formats, a high percentage of the respondents stated that they would recommend an online course format even if they did not prefer it to traditional instruction. They cited advantages of online instruction as delivering content that can be accessed at an individual pace, flexibility in when and where content is accessed, and lowered stress levels for working students with full and undependable schedules. They cited disadvantages as low interactions with the instructor or other students and concerns regarding the hardware and software. Students stated that they learned as much from accessing the online materials as they did from coming to the traditional class sessions, and most stated that they would take another online course.

It is clear that the success of online courses is contingent on a number of factors. The following sections will address a number of factors that are crucial in online delivery of content: (a) planning for online course delivery, (b) preparing online materials, (c) planning for course interactions, (d) delivering individual units of instruction, (d) testing and evaluating online student progress, (e) facilitating collaborative activities, and (f) preparing students to be online learners. The chapter will conclude with a case study of an online course offered in early childhood special education at the University of Kentucky and recommendations for instructors who are considering online delivery.

FACTORS THAT ARE CRUCIAL TO ONLINE DELIVERY

Planning for Online Course Delivery

While online delivery operates under a number of constraints, it also opens up a world of possibilities that do not exist in a traditional delivery format. Sufficient planning is the key to developing a successful online course. Instructors should bear in mind that the content should guide the type of online delivery in considering which components to develop.

Making course format and technology decisions. Prior to development, there are both student and course development factors that should be considered in the design of an online course (Schnorr, 1999). Student considerations include (a) who will be taking the course, (b) where students in the course live, (c) the resources that are available and accessible to students, (d) the technology skills of the students, (e) the availability of technical support, (f) whether students are full- or part-time, and (g) whether students also hold full-time jobs. Course development factors include (a) the type of course the instructor desires, (b) whether or not the content is suited to online delivery, (c) the type of format that matches the content, (d) the components the instructors plan to personally develop, (d) the availability of resources that may be needed, (e) who will develop time-consuming components, (f) copyright issues, (g) the availability of technology, and (h) costs and funding.

Once student and course development factors are determined, the instructor will want to consider the inclusion of a number of course components (Collins, Schuster, Ludlow, & Duff, 2002; Hains, Belland, Concelcao-Runlee, Santos, & Rothenberg, 2003; Ludlow et al., 2002; Meyen & Lian, 1997a, 1997b). Depending on the complexity of the course, components may include the following: (a) text-, audio-, and video-based online content; (b) links to Websites in the field, (d) electronic reserves of readings or other supplemental materials, (e) listservs or other group communications, (f) online forums or discussions, (g) electronic presentations, (h) introductory Website or page, (i) course syllabus, (j) course outline or table of contents, (k) list of resources, (l) evaluation forms, (m) course activities, (n) class roster and introductions to the instructor and students with optional pictures, (o) e-mail capability, and (p) quizzes.

In addition to these components, the course developer will want to determine whether delivery will be synchronous (available in real time), asynchronous (available at a delay) or a combination of the two (Collins et al., 2002). For example, the instructor may choose to post content in advance but have real-time chats to discuss content, or the instructor may choose to present content through live videostreaming but have students interact over time on a discussion board. Collins et al. (2002) recommended that the instructor also may wish to consider the posting of course "netiquette" (i.e., rules for participation that shows consideration for the instructor and other students in the course) for students prior to their participation, a course design that includes variety and emphasizes activity-based learning, teaching strategies that are appropriate to the content, course components that are presented in manageable segments, and the scheduling of adequate time for interactions with the content, the instructor, and other students.

All of these components may be addressed in a course management system (e.g., WebCT, BlackBoard). When Day and Sebastian (2002) interviewed six students enrolled in computer-enhanced courses in visual impairments through the

University of Utah and Utah State University, they found that the students had definite opinions about the delivery technology and the course management system. They liked EDNET (a distance education system that uses multiple technologies including interactive video and audio) because it provided them with the opportunity to view all of the class sites and to interact in real time; however, they viewed the time spent driving to EDNET sites and auditory problems with the microphones as drawbacks to this technology. The students found WebCT easy to use and liked the functions (e.g., gradebook, calendar, online readings, bulletin board postings, e-mail); however, they reported server issues in rural sites (e.g., static and interrupted services), lack of time to use all of the features of the system, and the need for assistance in using the system. The students also reported personal technology problems, such as having inadequate home equipment for printing materials and finding it necessary to access the course system at work or school instead of home. Because course design is crucial to the success of online learning, instructors should seek support from online professionals when designing a course format (Schnorr, 1999), regardless of whether or not they are using a course management system.

Determining team members and timelines. When professionals work together as a team to develop a course, Collins et al. (2002) recommended that the individual responsibilities of the team members should be clarified, and an entire semester should be devoted to planning and development. For example, Ludlow et al. (2002) developed a special education course using WebCT and found the following timelines to be necessary for each member of the development team: (a) instructor – 2 wks to prepare an outline and assemble the materials necessary to prepare the content and 4 wks to write six modules, (b) content assistant – 3 wks to contact and schedule course guests and to write case studies and comprehension questions, (c) technology assistant – 2 wks to prepare the basic course structure and 4 wks to program the Web formats, and (d) media producer – 2 wks to record audio- and video-segments and 4 wks to edit and digitize media files. The development process began in January and was completed in May.

Based on their experiences in developing 11 Web-based courses in special education at Michigan State University, Bruce and Hwang (2001) recommended that a year should be devoted to planning and developing an online course. Specifically, Bruce and Hwang found that the preparation time per lecture hour was 2 to 10 hrs for on-campus delivery and 5 to 23 hrs for online courses. When time constraints prohibit sufficient development time, Collins et al. (2002) recommended that a minimum of three class sessions should be posted prior to the time that students begin a course.

Ensuring accessibility for all students. One aspect of course development that is often overlooked is the issue of accessibility for persons with disabilities. The Americans with Disabilities Act of 1990, however, mandated that all private and public schools must ensure that their programs and services are accessible and usable by persons with disabilities (Kessler & Keefe, 1990). This law applies to the manner of instructional delivery as well as the accessibility of educational facilities. While students who take online courses in the comfort of their own homes are responsible for the learning environment, the moment that they tap into a Web-based course, the course developers become responsible for the accessibility of the course (e.g., availability of audio text through a sound card for persons who are blind or a visual alternative to audio for persons who are deaf). Developers may not realize that a number of standard practices result in inaccessible content for students.

As of 2002, 10% of the of students enrolled in the Department of Special Education and Rehabilitation at Utah State University had visual impairments. Cain and Merrill (2002) surveyed seven of these students regarding online accessibility. The sample included 3 students who were blind, 2 students who had some vision, and 2 students with declining visual ability who had experience with three types of delivery technology: (a) teleconferencing, (b) online chats using RealAudio, and (c) Rotor (a combination of chat and media streaming). The researchers found that preferences varied, with those with low vision preferring Rotor (which was not accessible with screen reading software) and those who were blind preferring teleconferencing. They also found that students made other accommodations, including the use of Windows software to change the type of font and to increase font size and contrast, the use of a large monitor, and the simultaneous use of one computer to take notes and another to participate in class using Rotor. The students stated that they felt more included during teleconferencing and noted that the telephone leveled the playing field since all students had to rely on their hearing instead of their vision. This investigation illustrates the need for selections of adaptations and accommodations based on individual preferences and the value of gathering feedback based on students' experiences.

Fichten, Asuncion, Varile, Fossey, and Simone (2002) conducted a series of investigations regarding online accessibility for students with disabilities. In the first investigation, they asked 12 postsecondary students with disabilities about the advantages and disadvantages of computer and adaptive computer technologies. The students listed advantages as the potential of computer technologies to access information, to provide communication for those who are deaf or have physical impairments, to organize neater work in a timely fashion, and to decrease dependency. Barriers included problems with spelling; the need for training, retraining, and assistance; attitudinal problems of acceptance in the classroom; the physical exertion of typing for those with physical impairments; and cost and compatibility of software and hardware.

In the second investigation, Fichten et al. interviewed 37 students with disabilities in higher education and 30 higher education personnel providing disability services. The majority stated that their disabilities affected school performance, half reported difficulty using a monitor and a mouse, and many reported problems using a keyboard, manipulating a disk, or using a printer. Again, funding for equipment was a problem, with larger institutions more likely to have larger numbers of students with disabilities and, thus, a greater availability of assistive or adaptive equipment. The researchers also found that students across various types of disabilities tended to share the same types of adaptations. For example, students with learning disabilities and students with visual impairments both used screen reading software. In addition, they found that a large number of students used mainstream adaptations (those used by persons without disabilities), such as dictation software, spell check programs, and scanners.

In the third investigation, Fichten et al. analyzed the responses to questionnaires received from 725 students with disabilities in higher education across Canada to determine the type of assistive or adaptive equipment used across disabilities. They found that the majority of the students used computers (e.g., research, e-mail, library access), and almost half needed computer adaptations. The barriers most often cited included purchase costs, availability, and maintenance/upgrade costs. The most popular adaptations were mainstream equipment, including spelling and grammar check programs, scanners, portable notetakers, and dictation software. Based on their findings, Fichten et al. recommended that persons with disabilities be included in the evaluation of courseware, that experts in accessibility be included in the design process, that course management systems be adopted that have built-in features for persons with disabilities, that free Web-based tools (e.g., Bobby for persons with visual impairments) be used to evaluate course accessibility, and that courses be designed with initial attention to accessibility rather than trying to retrofit them. Table 1 lists a number of adaptations for students across a variety of disabilities.

Preparing Materials for Online Delivery

Online instructors need to give consideration to the preparation of two types of online materials: (a) those that

Planning & Delivering

will be used to deliver instruction and (b) those that will be used to supplement instruction. The professional literature contains several examples of materials that have been placed online in the delivery of special education personnel preparation programs (Hains et al., 2003; Ludlow et al., 2002; Meyen & Lian, 1997a). Instructional materials have included (a) links to the field, (b) lecture notes and outlines, (c) exams and other assessments, (d) journal articles and book chapters, (e) PowerPoint presentations, (f) audio- or video-clips, and (d) glossaries. Supplemental materials have included (a) lesson schedules, (b) reading lists, (c) course syllabi, (e) listservs, (f) class rosters, and (g) e-mail addresses of the instructor and technical support personnel.

Preparing text-, audio-, and video-based instructional content. In delivering instruction, the general rule is to use formats and strategies that best convey the content. For example, Bruce and Hwang (2001) presented text-based content in special education that was supplemented with clip art, cartoons, brief film clips, audio stories, simulations, Website links, and guided activities. Text-based documents can be created in HTML (hypertext markup language) or PDF (portable document format) files (Ludlow et al., 2002). Because HTML may display differently across platforms and browsers, PDF may be a better choice (Schnorr, 1999). The use of a conversational tone in written text may be more appealing and effective in an online course than technical writing (Bruce & Hwang, 2001). Audio and video content should be presented in short, explicit clips – 20 to 30 s in duration for audio and 60 to 90 s in duration for video (Ludlow et al., 2002). In general, content that is videostreamed should follow the guidelines for video-based courses (e.g., satellite, interactive video), such as avoiding distracting jewelry or patterns; neutral colors, however, translate better to computer monitors than the brighter colors that are desirable on televised courses (Collins et al., 2002). In creating video for online courses, Schnorr (1999) suggested (a) using a tripod, (b) avoiding titles or zooms that take up storage space, (c) focusing on people and actions, (d) recording extra footage that can be edited at a later time, (e) flooding subject matter with light, (f) avoiding glare from windows and lights, and (g) using a high quality camera.

Designing the course Website. Particular attention should be given to the design of a course Website where content will be presented, whether or not a course management system is available or desired. In developing a course Website, Hains et al. (2003) made the following recommendations:

1. The appearance or layout should have a common look across pages with an identifying logo, text should be brief and written in active voice with no more than 60 words per line, the use of boldface and italics should be limited, bullets and organizers should guide the reader, and white space should be used liberally.

2. The navigational design should have a continuous site image on the screen with clear methods and limited options, easily accessed controls, and location information.

3. Hypertext features should include minimal scrolling with links to definitions of technical words and other Websites and with search engines available within the site.

Planning & Delivering

4. Legibility should address the visibility of materials on all screen sizes and minimal distracting flickering or flashing.
5. Graphic and video elements should be quick in delivering information, quick in loading, and short in duration (less than 2 min) with thumbnail views available of large graphics.
6. Electronic accessibility should include HTML text, large areas for pointing and clicking, contrast in color combinations, information delivered by means other than color, pop-up windows that can be turned off, accessible video (e.g, captions, audio descriptions, transcriptions), vocabulary within a range of reading levels, defined acronyms, and a separate page when a primary page cannot be made accessible.
7. E-mail contact information should be included for the site manager and the group responsible for the site.
8. All users should be encouraged to complete an evaluation of the site.

PLANNING COURSE INTERACTIONS

Because students do not have face-to-face interactions in a course that is fully online, it is necessary for online instruction to devote special planning on ways to increase interactions. Three types of interactions should be considered: (a) interactions between the student and the content, (b) interactions between the student and the instructor, and (c) interactions between the students and other students in the course. For an interaction to occur, there must be a two-way exchange of information. For example, the instructor or a class member may communicate with a student, but an interaction does not occur until the student responds. Likewise, content may communicate information to a student, but an interaction does not occur until the student responds to the content in some way. Interactions may be synchronous and asynchronous. Options for interactions include face-to-face audio or video components, online chats, electronic bulletin boards, moderated discussion boards, and shared whiteboards available through the use of native Web capabilities, commercial groupware products, programming tools with templates, and hybrid designs (Gilbert & Moore, 1998).

Student-content interactions. Student-content interactions can be facilitated by presenting interesting content in a conversational tone and encouraging active participation through course components, such as video, threaded discussions, e-mail, chat, quizzes, reading summaries, and in-depth papers and projects (Bruce & Hwang, 2001). Course activities require students to demonstrate understanding. It is helpful if the instructor provides guidance through clear directions, structures activities to allow students to express their viewpoints, and monitors progress through activity response forms (Meyen & Lian, 1997a, 1997b). It is also important that the rules for content interactions are clear from the beginning of the course. This includes having clear timelines for projects, set times for midterm and final exams, and target dates for completing units; these rules allow students to pace themselves

(Meyen & Lian, 1997b). Students also can benefit by knowing the value that is placed on course projects and that they will not be held responsible for optional content (Meyen & Lian, 1997b). When activities are of exceptional quality, sharing them with other students (e.g., listserv) may be desirable (Meyen & Lian, 1997b); this can be reinforcing to the student who completed the assignment and may serve as a model for other students who are having difficulty.

Student-instructor interactions. Student-instructor interactions also are key components in the online learning process. Instructors can create a more personal environment that is conducive to these interactions early in the course by sending a welcome message to students via a listserv or by placing photographs of the instructor and the students online (Meyen & Lian, 1997b). Subsequent interactions between the instructor and the students can be private (e.g., e-mail or phone) or public (listserv or group announcements on course bulletin board) (Meyen & Lian, 1997a). Regardless of the means of communication, instructors should attempt daily communication. Even the best online format does not eliminate the need for mentoring (Bruce & Hwang, 2001), so online instructors should expect to receive comments and questions from students and should be prepared to give detailed feedback as questions arise or as assignments are completed.

Because e-mail from students can become both cumbersome and even overwhelming, especially if enrollment numbers are high, the instructor will want to allot scheduled time for responding to e-mail in a timely fashion (Collins et al., 2002), realizing that quick replies are of particular value for students who have trouble staying on task (Meyen & Lian, 1997b). Optimally, e-mail should be checked on a regular schedule 2 to 3 times per day during the week (e.g., morning, afternoon, evening) and once per day on weekends (Meyen & Lian, 1997b). Providing virtual hours when students can expect the instructor to be online is an excellent practice; high enrollment, however, may create the need for both day and night virtual hours (Schnorr, 1999). Bruce and Hwang (2001) illustrated the necessity of allowing adequate time for e-mail when they reported higher levels of student-instructor e-mail messages in online classes than in traditional on-campus classes. In their experience, on-campus students sent 1 to 9 e-mail messages to the instructor per semester while online students sent 6 to 24. They noted that a 3-hr course may require 4 to 5 hrs of time to be allotted to e-mail per week.

Strategies for organizing incoming e-mail may be helpful to instructors (Meyen & Lian, 1997b). Creating individual student e-mail files and reviewing messages in the order in which they arrive decrease confusion. When instructors teach more than one online course at a time, separate e-mail accounts can be developed for each course.

Instructors can avoid some of the problems inherent to instructor-student interactions in online courses by setting the stage for communication. The instructor can place a limit on the length of student e-mail messages, let students know in advance their pattern for responding to e-mail, clarify when phone or office visits are appropriate, make clear whether casual e-mail messages are acceptable, and provide circumstances when students should contact the instructor for technological assistance (Meyen & Lian, 1997b). Several strategies may be useful in responding to student e-mail. These include using a conversational style; cutting surplus words; using words that are short, familiar, and precise; and limiting the use of jargon (Meyen & Lian, 1997b).

Instructors also may wish to consider several strategies in using e-mail to provide feedback on assignments (Meyen & Lian, 1997b). Providing positive comments gives students reinforcement. When initially providing corrective feedback, instructors can share experiences that illustrate better responses in lieu of being critical; feedback on assignments can become more analytical over time. Posing questions for clarification of assignments can facilitate self-reflection. When instructors fail to hear from students or hear from them sporadically, brief inquiries from the instructor can keep them on track with timelines; when students are falling behind or the quality of their work is declining, daily e-mail contact can be encouraging.

Student-student interactions. Student-student interactions build a sense of community and belonging in online courses where face-to-face interactions do not occur. Using open forums, such as during synchronous online chats or asynchronous discussions boards, can facilitate student-student interactions. These types of forums give shy students the opportunity to participate, and extended postings allow time for reflections on points made during discussions (Bruce & Hwang, 2001). The design of asynchronous discussions is key to their success in facilitating student-student interactions, as illustrated by the following recommendations (Lawrence,

1996-97). The initial greeting and the discussion item should be posted with careful wording in brief and simple terms. Discussion items that are multi-faceted will result in disjointed dialogue or no dialogue at all. Items that are a bit controversial are desirable because they incite discussion. Once the discussion is under way, the instructor may need to steer the discussion and model appropriate responses. In some cases, the instructor may need to contact a student privately. For example, private messages may be used to encourage participation in the discussion, to comment on posted responses that are too long, or to encourage responses to other students. Discussions should have specific timelines (e.g., 2 wks).

While the instructor monitors most chats and discussions, it may be advantageous to create an open forum in which the instructor does not participate and there are no clear guidelines. This will allow students the opportunity to get to know each other, to provide support for each other, and to build a social and professional community beyond the confines on the course.

Planning Individual Units of Instruction

In addition to planning the overall course structure and format, instructors of online courses must plan for individual units of instruction. The advantage to online learning over traditional on-campus class sessions is that students can choose to complete an online topic in a single session or to break it up over time (Bruce & Hwang, 2001). The variety of options for presenting content can be illustrated by the following examples of online courses in special education.

In designing the delivery of course content, Ludlow et al. (2002) divided a course into six modules. The first module provided an introduction to the course. Four content modules followed this. The final module provided a summary of the content presented in the course. Each module consisted of 10 to 20 pages of single spaced text written in conversational style, photographs, audio clips, video segments, and linked reading and Websites. In addition, students completed learning activities that consisted of experiential activities, comprehension checks, and case study analysis.

In two special education courses, Meyen and Lian (1997a) presented content through audiostreamed lectures, images of the instructor teaching onscreen, and graphics of the discussion topic. Students could navigate forward or backward in each lesson and could see key points listed at the top of the screen. The courses also contained a glossary of terms, a text outline of each lecture, embedded activities, progress assessments with feedback, scheduled breaks, concluding quizzes on topics, and opportunities for review.

Planning for Testing and Evaluation

Testing allows online instructors to assess the learning that occurs in their courses, while evaluation allows online instructors to determine the effectiveness of their delivery format and to identify areas where improvements can be made.

Testing. Frequent quizzes and graded activities provide feedback on learning for both the students and the instructor (Meyen & Lian, 1997b). Cheating may be of particular concern in online courses (Meyen & Lian, 1997b) since students are at a distance from the instructor. To alleviate this concern, the instructor may wish to have students come to campus or monitored sites to take exams. Another option is to use open-ended exams that require understanding and application of concepts rather than recall and memorization. A third option is to create windows of time in which quizzes and exams are accessible, such as creating a set number of minutes or days in which an assessment is available to be completed (e.g., exam posted for two days with 20 min of time allotted to complete it once the exam is accessed). It should be noted that online courses may have high enrollments; thus, grading support (e.g., grading assistant, electronic grading) may be needed (Schnorr, 1999).

Conducting evaluations. It is important that students separate the evaluation of the course instructor, the design of the format to deliver content, and the delivery technology. Thus, course evaluations need to have items that address each of these measures. Course evaluation can be conducted through various formats. For example, Ludlow et al. (2002) used four measures: (a) audit trails, (b) an online survey, (c) focus group sessions, and (d) a follow-up questionnaire. Using more than one type of evaluation measure increases reliability on the responses. For example, Day and Sebastian (2002) described a case in which an online student reported that he or she did not receive enough contact from the instructor, while data showed that the instructor sent 41 personal e-mail messages to the student in addition to making numerous phone calls.

PREPARING STUDENTS FOR ONLINE LEARNING

Hains et al. (2003) noted that a digital divide is still evident in higher education. This is especially true in online courses, where younger, traditional students may enroll because they are attracted to and fluent with technology, while older, non-traditional students may enroll due to convenience although they have minimal and non-existing familiarity with online technology (Day & Sebastian, 2002). Since students with better technology skills report more positive outcomes from online coursework (Day & Sebastian, 2002), instructors should facilitate familiarity with course technology. The first step in working with students new to online delivery is providing them with access to the technology they will need; this can be followed by training in the use of the system, software, or materials through online exercises (Day & Sebastian, 2002). For example, one way in which an instructor can increase computer comfort is to create a "treasure hunt" in which students search for information on various Websites (Hains et al., 2003).

It is likely that round-the-clock support will be needed for online students (Day & Sebastian, 2002). This is because technological problems can occur at any given time, and students will be accessing coursework at various times that work into their schedules. Technological problems may occur due to difficulty in accessing the course, differing preferences for specific software, the availability of an overabundance of Web tools that may prove confusing to students, and varying levels of student expertise (Murphy & Cifuentes, 2001).

To solve these problems, students must be willing to plunge in by contacting mentors or tutors, asking questions, sharing tips with other students, and consulting resource tools (e.g, online software manuals, library resources, printed guidelines, and tutorials) (Murphy & Cifuentes, 2001). The ultimate responsibility for securing support rests with the student rather than the instructor. For example, Day and Sebastian (2002) described a case in which a student who claimed to need more technological support failed to take action, even when told by the instructor to phone or e-mail a technology support person who would travel free of charge to the student's work or home site.

Another way to get off to a good start in an online course is to set course rules or guidelines for students. For example, students may benefit from a list of suggestions for communicating via e-mail (e.g., using first person language, being complete and specific, being redundant, asking for feedback on how a message is received, describing behavior without evaluating or interpreting it) (Meyen & Lian, 1997b).

Working Collaboratively

As mentioned earlier, student-to-student interactions are desirable in online coursework because they build a sense of community when students cannot meet face-to-face, allowing students to share information with each other. In addition to online chats and discussions, collaborative activities allow students to interact with both each other and the course content. It may be best to postpone collaborative projects until the course is underway and students are familiar with the technology (Meyen & Lian, 1997b). Once a collaborative project is assigned, the instructor will want to provide detailed information about the assignment, the timelines for completion, and the roles of team members, in addition to monitoring progress of groups as they work together (Meyen & Lian, 1997b). Communication between students can occur through facsimiles, telephone, e-mail, or face-to-face contact (Meyen & Lian, 1997b).

As an example, Murphy and Cifuentes (2001) required two collaborative projects from 13 graduate students who were new to online instruction and were enrolled in a course on telecommunications in education. The projects consisted of discussing course readings on a discussion board and conducting research in small groups that resulted in a written chapter centered on a theme. Students interacted with each other through three spaced face-to-face meetings, asynchronous discussions on an online discussion board, and synchronous online chats. Murphy and Cifuentes identified several problems that made online collaboration difficult. These included difficulty in gaining online consensus, misunderstanding of each other's roles, technology problems for novices, and differences in expectations, culture, knowledge, motivations, and preferences. To address these problems, Murphy and Cifuentes scheduled an orientation to develop group learning contracts and profiles of students that included their learning styles, personalities, expertise, and perspectives. This allowed students to learn to respect each other and learn from each other. Based on their experience, Murphy and Cifuentes recommended that a balance needs to be established between structure and dialogue, ensuring that students have a high degree of participation and interactivity and realizing that group activities build a sense of community in online courses.

A CASE STUDY OF AN ONLINE COURSE AT THE UNIVERSITY OF KENTUCKY

Most of the factors discussed so far in this chapter can be illustrated by the experiences of an instructor at the University of Kentucky who taught an online course for the first time. An overview of the course and the instructor's experience are detailed in the following sections.

Description of Course

The course was a freshman level course that is an introduction to early childhood special education. The course dealt with development, curriculum, and types of childcare offered in early childhood (i.e., birth to 8 yrs of age).

Description of Participants

Instructor. Two doctoral students in early childhood special education shared responsibilities in teaching the course. One had the role of instructor, and the other served as an assistant. They shared the tasks of leading online discussions, answering e-mail, and grading class assignments.

Students. There were 29 students enrolled in the course, with 28 completing. All students were undergraduates, with 21 seniors, 7 juniors, and 1 sophomore. Students came from a variety of disciplines, including Family Studies, Early Childhood, and Speech/Language Pathology.

Planning & Delivering

COURSE DEVELOPMENT

The course had been developed by a professor and was used as a training ground for doctoral students to receive experience in teaching via distance education. The course included 15 modules, and half delivered instruction through video. The lead instructor was given responsibility for teaching the course 3 days before classes started. During this time, she had to find and participate in training for Blackboard 6, update the course due dates, and make sure the online portion of the course was ready for delivery.

COURSE STRUCTURE

The delivery method was hybrid, consisting of videos that aired on a public educational channel and asynchronous online delivery through Blackboard 6.

Course materials. The students were required to buy a textbook, as well as to visit relevant Websites and to download a chapter from a Website.

Assignments. Assignments for the course included reviewing Websites (both guided and independent), answering questions about case studies, interviewing families and professionals in the field, and writing a research paper on a relevant topic.

Interactions. Students interacted with each other and both instructors through a discussion board. Since there were so many students in the class, the students were placed into 1 of 4 groups. Half of the discussion questions were answered by all students, and half were answered by 1 of 4 groups. On average, a student responded to a discussion board question three times per month during the semester. Each student was required to post to the discussion board twice to earn all possible points, with the exception of the first and last discussion board post. While most students interacted with the instructors through e-mail, there were a minimal number of phone calls and face-to-face meetings with a few students to solve specific problems.

Assessment. Students grades were determined by points allotted for the following assignments: (a) review of four Websites, (b) interviews with two preschool teachers, (c) interviews with families of preschool students, (d) questions related to case studies, (e) discussion board participation, and (f) a research paper on a relevant topic.

Reflection

The course provided excellent hands-on experience in delivering online content, and the course instructor learned some valuable lessons that will enable her to refine future online delivery. These include the following:

1. The course would have been more effective with fewer students.

2. The instructors should have had more preparation time.

3. The assignments should have been rearranged to allow more time for thoughtful grading and to allow students to know their point standings throughout the course.

4. All students should have been made aware that they

Planning & Delivering

should use e-mail accounts within the course management system.

5. Supplemental materials (e.g., textbook) should have been available at the beginning of the class.
6. The instructor should have made clear that Word was the format preferred for submitting assignments since some documents could not be opened.
7. Due date times should not have been midnight because this created confusion over which day assignments were due.
8. There should have been ways to assess student knowledge (e.g., tests).other than through written products
9. Students should have been assessed on their ability to use the functions of the course management system.
10. Netiquette should have been addressed through an assignment or posted rules.

While this course gave the instructor the opportunity to practice what she had learned in previous coursework on distance education management and delivery, all first-time online instructors should view online delivery as a dynamic process and reflect on ways to improve future course delivery based on their experiences. In addition, having a knowledgeable mentor can aid first-time instructors in refining techniques as they progress through the course. For example, there were issues that arose in this course in which the instructor was able to turn to experienced distance educators for assistance. These included dealing with plagiarism, late assignments, and unresponsive students.

RECOMMENDATIONS

Based on the professional literature and their experiences as online instructors, the authors recommend that online instructors give careful consideration to planning and delivery. In planning, it is important to allot sufficient time for developing the course and becoming familiar with the technology and the course management system. In addition, all courses should be planned with accessibility for all students in mind. Also, in selecting technology, the content should be the driving force. In delivery, instructors should allot sufficient time for course management, know what resource are available when technology or student problems arise, and give careful thought to multiple ways to assess students' understanding of the content.

REFERENCES

Beard, L. A., & Harper, C. (2002). Student perceptions of online versus on campus instruction. *Education, 122,* 658-664.

Bruce, S. M., & Hwang, T. (2001). Web-based teacher preparation in visual impairment: Course development, teaching, learning, and quality assurance. *Journal of Visual Impairment & Blindness, 95,* 609-623.

Cain, H. M., & Merrill, Z. (2001). Distance education for master's students with visual impairments: Technology and support. *Journal of Visual Impairment & Blindness, 95,* online.

Collins, B. C., Schuster, J. W., Ludlow, B. L., & Duff, M. (2002). Planning and delivery of online coursework in special education.

Teacher Education and Special Education, 25, 171-186.

Day, J. N., & Sebastian, J. P. (2002). Preparing vision specialists at a distance: A qualitative study on computer-enhanced learning. *Journal of Vision & Blindness, 96,* 796-808.

Fichten, C. S., Asuncion, J. V., Varile, M., Fossey, M., & Simone, C. (2000). Address to educational and instructional computer technologies for post-secondary students with disabilities; Lessons from three empirical studies. *Journal of Educational Media, 25,* 179-202.

Gilbert, L., & Moore, D. R. (1998). Building interactivity into Web course: Tools for social and instructional interaction. *Education Technology, 38,* 29-35.

Hains, A. H., Belland, J., Concelcao-Runlee, S., Santos, R. M., & Rothenberg, D. (2003). Instructional technology and personnel preparation. *Topics in early Childhood Education, 20,* 132-144.

Kessler, D., & Keefe, B. (1999). Going the distance. *American School and University, 71*(11), 44-47.

Knapczyk, D., Rodes, P., Haejin, C., & Chapman, C. (1999). Collaborative teacher education in off-campus rural communities. *Rural Special Education Quarterly, 28*(3/4), 36-44.

Lawrence, B. H. (1996-97). Online course delivery: Issues of faculty development. *Education technology Systems, 25,* 127-131.

Ludlow, B. L., Foshay, J. D., Brannan, S. A., Duff, M. C., & Dennison, K. E. I2002). Updating knowledge and skills of practitioners in rural areas: A Web-based model. *Rural Special Education Quarterly, 21*(2), 33-44.

Meyen, E. L., & Lian, C. H. T. (1997a). Developing online instruction: One model. *Autism and Other Developmental Disabilities, 12,* 159-166.

Meyen, E. L. , & Lian, C. H. T. (1997b). Teaching online courses. *Focus on Autism and Other Developmental Disabilities, 12,* 166-175.

Murphy, K L., & Cifuentes, L. (2001). Using Web tools, collaborating, and learning online. *Distance Education, 22,* 285-305.

Schnorr, J. M. (1999). Developing and using technology for

course delivery. *Teacher Education and Special Education, 22,* 114-122.

TABLE 1

Potential Assistive and Adaptive Computer Equipment for Students with Disabilities Based on a List Developed by Fichten et al. (2002).

	Assistive and Adaptive Computer Equipment for Students with Disabilities		
Students Who Are Blind	Voice synthesizer	Screen reader	Document reader
	Scanner hardware and software	Text-based browser and e-mail software	Voice-activated mouse control
	Portable Braille 'n' Speak or Type 'n' Speak	Braille translation software and pointer	
Students Who Have Low Vision	Voice screen and document read	Scanner hardware and software	Large monitor with visors or masks
	Portable Type 'n' Speak	Screen with magnification	CD-Rom encyclopedia
	Display control for zooming, changing font size, and increasing contrast	Document management software	Voice-controlled menus and toolbar

Assistive and Adaptive Computer Equipment for Students with Disabilities

Students Who Have Hearing Impairments	CD-Rom encyclopedia	Captioning. e-mail, and chat programs	Portable C-Note system
	Spell check and grammar check software	Control display with visual flash	
Students Who Have Motor Impairments	Keyboard with sticky keys	One-handed typing software	Write rests and key repeat adjustments
	Voice-controlled menus and toolbar	Voice-recognition software	Scanner hardware and software
	Monitor and LCD projector	Word prediction software and e-mail	AlphaSmart notetaker
	Ergonomic work station, desk, chair, and carrel	Joystick, trackball, touch pad, or ergonomic mouse	Input devices (e.g., sip and puff switch, mouth wand)
	Adjustable keyboard, monitor, computer, and document stand	Portable language master and spell checker	Keyguard splint

Chapter 3

Providing Technology Support to Learners

Katherine J. Mitchem
California University of Pennsylvania

Online instruction has become an attractive option for delivering coursework to individuals who, for a variety of reasons including family responsibilities or geographical and time constraints, might otherwise not have access to further education. Advantages of online instruction include flexibility of use, convenient access, potential to reach large numbers of students across a wide geographic area, and the opportunity for synchronous and asynchronous interaction among remote students. Of course, this access does not come without its own set of challenges. Many distance learners are older, non-traditional students whose prior experience in education did not require internet access, technological competence, and web-mediated participation. This somewhat paradoxical situation—the audience most likely to benefit from online instruction is the audience least likely to have the technology expertise to access it—prompts the question how best to provide technology support to learners. This chapter is divided into three sections addressing the following questions: (a) What information do learners need to be successful? (b) What support services should institutions provide? and (c) What role should the instructor play in providing technology support?

WHAT INFORMATION DO DISTANCE LEARNERS NEED TO BE SUCCESSFUL?

For students to be successful in a virtual environment, they must be able to access the information and materials provided through the course as well as have the skills and confidence to interact with those materials. In general, students need to learn what hardware and software is necessary to be able to access all course information as well as how to navigate the course, access materials, interact with others, and submit assignments. Some students may need tips on how to become self-directed and self-regulated learners. It is essential for students to know how to access technical support or a help line (both online, email and telephone support) as well as how to take a problem-solving approach to the inevitable technology and connectivity problems. A more detailed discussion of the required hardware and software, learner skills, and policies and procedures to facilitate online learning follows.

What Hardware and Software Is Needed?

Learners need to know what hardware and software is required to participate so that they are able to make any necessary upgrades before the course begins. Students need information on the specifications of the operating system required to work with the web platform for both PC and Mac users. In addition, minimum processing speed for the CPU and minimum megabytes of internal memory (RAM) should be provided. Currently, most courses require a minimum of 300 MHZ processing speed and 64 MB of internal memory (RAM). These numbers vary depending on the level, sophistication, and type of technologies incorporated in the course. Students also need reliable access to the internet with reasonable bandwidth and speed, no firewall restrictions, and relatively few interruptions in service due to high traffic. A 56 kbps telephone modem is adequate; slower modems (14.4, 26.8 kbps) will provide poor access. For online courses delivered through webcasting (Ludlow, 2005), a separate phone line is necessary to participate during webcasts. Both cable/satellite modems and digital subscriber lines at 128 kbps provide excellent access and allow simultaneous phone contact. If the course requires the use of voice boards or voice chats, students will need a microphone.

How To Be a Successful Distance Learner

Online education is student-centered, with instructors defining the goals and facilitating the learning process and students discovering content on their own (Kearsley, 2000). Independent and collaborative learning are stressed, creating difficulties for students who are used to simply taking notes and answering questions in class (Ko & Rossen, 2004). The literature on student skills needed to take advantage of online learning is limited (Collins, Shuster, & Grisham-Brown, 1999; Menlove & Lignugaris/ Kraft, 2004) and typically focuses on performance of students in distance education courses compared to student performance in traditional campus-based courses. For example, Jordan et al. (2004) compared instruction provided in a traditional course to instruction provided in a web-enhanced course. Students completed a self-assessment on their skills and knowledge both in content and technology and also were asked to report their confidence in technology skills. While the authors found improvement in content skills in both groups and in technology skills in the web-enhanced group, they indicate the need for further research into variables related to student satisfaction with courses, such as belonging and connectedness.

The new demands placed on students—both for self-directed, independent work as well as competence across a range of technologies—can lead to both higher drop-out rates than in traditional classes as well as increased demands on instructors. Biner, Dean, and Mellinger (1994) argued that high levels of student satisfaction are important because they may contribute both to lower student attrition and greater student learning in the course. Other authors have identified student frustration with technology as one of the main problems interfering with the amount of energy and effort that students can dedicate to learning content (Ludlow & Duff, 1998). Menlove and Lignugaris/ Kraft (2004) and other authors (Hora & Kling, in press) have suggested that even when websites and course WebPages are developed to address problems associated with distributing course materials and communicating with instructors, the lack of technology skills needed to access the materials may create new barriers for some students. In online instruction, even a relatively simple task, such as asking the instructor a question, may be dependent on the student's technology skills. Those students lacking these basic technology skills can quickly become frustrated or disenfranchised, and this may lead them to perform poorly or even drop the course. Learners need to be supported through the often ample opportunities to experience both system failures as well as operator errors so the availability of training and support for online learners is critical (Owsten, 1997; Whitworth, 1999).

Menlove and Lignugaris/Kraft (2004) identified the following primary and secondary core skills required for success. Primary core technology skills are those needed to access information on course web pages and include word-processing and emailing using an Internet browser. Secondary core skills are those specific to the web design and instructional activities in the program's electronic delivery system and may include using Acrobat Reader, downloading from the Internet (to obtain the Acrobat Reader software), installing the software and plug-ins, reading academic text, and annotating readings on computer screen (Menlove & Lignugaris/Kraft, 2004, p.23).

In addition to having technology skills, online students also need another set of skills that are related to the student-centered nature of online learning. Online learning is not an easier way to get an education so students should plan to spend between 4 and 15 hrs per week per course (University of Illinois, n.d.) and should believe that high quality learning can take place without going to a traditional classroom. Students will be expected to work with others in completing projects, use critical thinking and decision-making skills, think ideas through before responding, and be able to use technology properly. Students must be self-directed and self-disciplined if they are to stay on top of assignments and readings and move through the course at an appropriate pace. Without the physical presence of the instructor to prompt procrastinators to turn in assignments on a particular date, some students may avoid task completion and fall so far behind in the course that they are unable to complete it. In addition, it is critical that students are comfortable with expressing themselves in writing as this may be the main form of communication both with the instructor as well as with fellow students.

In typical classroom instruction, instructors are accustomed to gauging students' understanding, frustration, boredom, etc., with visible indicators such as facial expressions, absence, and off-task behaviors. In the online classroom, these non-verbal gestures may not be identified as quickly so students need to be willing to voice their questions and concerns as soon as problems arise (Varvel, 2001).

Policies and Procedures for Online Learning

To help foster the development of organizational skills and independence in learners, detailed directions should be provided to help them prepare for online learning by making sure that they have the required hardware and software to access and interact with course materials. For example, Ludlow (2005) provides a web access letter to all students prior to the start of the course that provides detailed directions to students about the course schedule, technical assistance and support, and general preparations to make before class begins. Experienced instructors advise learners who are somewhat intimidated by technology to identify a family member, peer, or high school student to assist them with technology demands, such as downloading software and installing the browser version needed for optimal interaction with course materials and tools. In their study on preparing distance education students to succeed, Menlove and Lignugaris/Kraft (2004) noted that students who rated themselves low in technology skills but who did well in the course indicated that they had sons, daughters, husbands, or friends who helped them with the technology demands in the course. In addition to directions to prepare learners for a successful experience, detailed policies and procedures that address problems that may arise during the course are also essential. These should describe what students should do if they have difficulty accessing an on-line chat, lose work because of computer problems or a server outage, or are unable to submit work because of heavy online traffic.

WHAT SUPPORT SERVICES SHOULD INSTITUTIONS PROVIDE?

If faculty and students are to be successful in the online environment, they will need support in learning to use the technology as well as in troubleshooting problems that may arise (Collins, Shuster, Ludlow, & Duff, 2002). Institutions should provide support services to students that include support from entrance into an online education program to its completion. Support systems for faculty must accommodate the varied perspectives that instructors in online environments may bring to the table. Instructors who have experienced prior success with technology-mediated instruction will need a different type of technical assistance and support than faculty who are being encouraged to teach online for the first time. Institutional support systems must be designed to address the needs of both types of faculty member. These issues are addressed in more detail in Chapter 14 (Collins & Baird, this text).

Support systems must be designed to serve student learning outcomes as the effectiveness of such support affects the quality of the online experience for both faculty and students (Council for Higher Education Accreditation, 2000). Creating a technology infrastructure that supports web-based instruction and minimizes the likelihood of problems is an essential aspect of institutional support. Specific examples of what this infrastructure might contain include (a) access to an orientation session or module specific to the course management system used by the institution, (b) accessible help desk service available online or toll free, (c) frequently asked question (FAQ) lists to address common concerns and problems, and (d) accessibility provisions for individuals with disabilities.

Orientation Session or Module Specific to Course Management System

Many institutions adopt a specific course management system, such as WebCT, Blackboard, or TopClass, for the integrated set of tools that make instruction easier for faculty and students. This allows faculty to focus on development of content, technical personnel to focus on network maintenance and troubleshooting, and students to focus on learning the content rather than navigating unfamiliar web sites. If an institution has adopted a course management system, it makes sense to have an institution-sponsored module that orients students to aspects of the management system common to all courses, such as access to content and technical support, features of the system, and specific tips to facilitate a successful online learning experience. Pennsylvania State University (1998), for example, provides an orientation for all students that introduces them to the formats typically used for online teaching and learning. Other institutions may provide a group session on campus to demonstrate the technology and provide guided practice with technical support in a computer lab. Still other instructors may use an initial face-to-face session to introduce and demonstrate online learning formats (Menlove & Lignugaris/Kraft, 2004).

Some institutions provide a guide with simple, clearly written, and detailed directions for every aspect of an online course. Examples of these can be found at www.online.usu.edu (Utah State University), www.worldcampus.psu.edu (Pennsylvania State University), and www.cvc2.org (California Virtual Campus). To ensure broad availability and accessibility, these directions should be provided as online modules, print manuals, or pre-packaged videotapes (Collins et al., 2002). In addition to guides made available by institutions, some course management systems provide help function links and a user's guide from the company website. It is helpful to provide learners a link to this website from the course site.

ACCESSIBLE HELP DESK SERVICE AVAILABLE ONLINE OR TOLL FREE

Those students who typically access online courses often do so because of the convenience of access and the flexibility of scheduling that this type of instruction provides. Thus, these students need access to high quality technical support services at times that they are able to work on the course. It is more cost effective for institutions to provide technical assistance that is available 24/7 than for individual programs to do so. Most colleges and universities have a help desk staffed by personnel from academic computing services to support online and distant students. To avoid long-distance charges for students, a toll-free phone number is desirable. Unfortunately, Bates (2000) indicated that many institutions rely on a "one size fits all" approach to technical support rather than having support personnel trained to provide assistance ranging from simplified directions to beginners to expert consultation and resources for more advanced students.

FAQ LISTS TO ADDRESS COMMON CONCERNS AND PROBLEMS

An efficient method of providing written guidelines for identifying and resolving the most common problems that students face is to incorporate a frequently asked question (FAQ) section in the web course management system. When students encounter technology problems, instructors can foster the development of student problem solving skills by encouraging them to use "three before me" (unknown). They should check their orientation materials or guide first, the FAQ list second, and the help desk or technical assistance center third before contacting the instructor.

ACCESSIBILITY PROVISIONS FOR INDIVIDUALS WITH DISABILITIES

Since one of the primary purposes of online education is to offer students "learning anytime, anywhere," online resources should be designed to afford students with disabilities maximum opportunity to access resources without the need for outside assistance. Many government and educational institutions now require that all web pages follow accessibility guidelines established by the World Wide Web Consortium (W3C) (Foley & Regan, 2002). (See also http://www.w3.org/TR/WCAG10/ for W3C guidelines to accessible web content). One of the most important considerations in designing accessible websites is the provision of meaningful alternatives for content that may be inaccessible to some students. For example, the Accessibility in Distance Education website (http://www.umuc.edu/ade/howto/index.html) provides the following examples of meaningful alternatives:

- Captions added to a video for students who are hearing-impaired
- An HTML version of a Flash animation for students who are blind or who cannot navigate in Flash

This site, along with others (see Chapter 2 by Collins & Galyon-Keramidas, this text) provides examples of how to design web content to ensure accessibility to all students.

WHAT IS THE ROLE OF THE INSTRUCTOR IN TECHNOLOGY SUPPORT?

SELECTING TECHNOLOGY FORMATS THAT ARE ACCESSIBLE TO SPECIFIC LEARNERS WITH RESPECT TO BANDWIDTH AND CONNECTION SPEED

In selecting technology formats, instructors need to ensure that these are convenient for learners' use (Gibb & Egan, 2001), affordable, reliable, and simple to use (Ludlow & Duff, 2001). One strategy that can be used in selecting technology formats for online instruction is to determine "level of lowest common technologies (LCT)" (Simonson, Smaldino, Albright, & Zvacek, 2003, p. 108) through the use of a survey to identify the technologies available to learners. Instructors wishing to incorporate video clips and graphics into their courses should take into consideration that many of the learners may be using a telephone modem to download information from the web. For example, Pindiprolu, Peck Peterson, Rule, & Lignugaris-Kraft (2003) noted a number of student concerns in their study comparing three different web-mediated experiential strategies. When students accessed the course from off-campus, low bandwidth resulted in them being logged out while working; limited computer memory or lack of proper software resulted in video clips not playing; and restricted capacity of the chat tool to display information required students to take notes by hand.

Another strategy to ensure accessibility to content is to prescribe the minimum computer and telecommunication capability required for students to enroll in the course. Instructors should inform learners of the minimum hardware requirements in advance to allow students to upgrade their own equipment or to identify access to alternative equipment meeting those requirements. In a similar fashion, learning which internet browsers facilitate seamless access to the course and its components can reduce challenges faced by students when they first enter the online environment. Instructors should provide students a link and directions on how to download software required in the course that the students may not have.

Orientation Session or Module Specific to Course Formats

Online instructors have used a variety of activities to support learners in their initial online learning experiences and to minimize some of the challenges arising from the combination of having to learn new content delivered via a new format that requires additional technological competencies. Ludlow and Duff (1998) emphasized the importance of providing technical support for students so that they can focus time and energy on learning content rather than on mastering technical aspects of the course. Providing an initial face-to-face orientation that includes teaching students the needed technology skills prior to or at the beginning of the course as well as orienting students to the different features of the course format and course management system is one option for learners able to travel to a more central location (Menlove & Lignugaris/Kraft, 2004). This, however, may not be an option for programs that serve large geographic areas that cross state lines and national borders.

At West Virginia University, for example, video streaming technology is used to offer courses in low incidence special education across the United States and in several international locations (Ludlow & Duff, 2002). At the beginning of these courses, learners complete an orientation module that exposes them to the technologies they will be using in the course. These include how to access and use portable document format (PDF) files, text discussion board and chats, voice board chats, the course calendar, webcasts/media files, content module activities, and course mail. For novice online learners, the orientation module includes practice exercises to orient new students to the course management system (in this case, WebCT) before required activities are due and to provide the instructor with useful information about the students and their technology competencies. Veteran students complete only those practice exercises designed to provide the instructor with information about the students and any new technologies that will be introduced in this particular course.

Detailed Directions for All Online Activities

Instructors need to have clearly written instructions for all aspects of the course. These must be posted in a visible location, be easy to find, and provide step-by-step directions to students. Some instructors provide these directions, illustrated by screen captures, to explain formats and routines specific to a particular course. Directions should address topics, such as configuring the computer and browser, using all required course formats, and completing activities and requirements online. Instructors should make these accessible in variety of formats (printed learner packet, CD-Rom or DVD with packet, or a link to a PDF file that students can download). It is a good idea to have a novice test these step-by-step directions to make sure there are no glitches beforehand and to reduce the number of panicked cries for help from students once the course is underway.

Instructors also need to set and maintain timelines for completing assignments as well as clear expectations regarding deadlines for students. One suggestion is to require students to contact the instructor beforehand if, for some reason, they cannot complete the work by the deadline. Shift the responsibility to the student so that she or he contacts you with a plan to complete and attach points to this. This may assist in helping to develop self-directed learners rather than having students become dependent on the instructor to prompt and remind them to turn in assignments. Encourage students to save copies of work offline to prevent losing work prior to submission.

Encouragement or Incentives for Learners to Learn Technology

Instructors should make themselves accessible at frequent and predictable intervals to ensure that they are aware of and able to respond to problems or concerns that students have. Experienced online instructors recommend checking mail messages and discussion postings at least three times per day (morning, noon, and evening, including weekends) to ensure a timely response especially during the first few weeks of the course. After students have become more comfortable with the formats, instructors may reduce the amount of time spent online (Collins et al., 2002).

To encourage learners to explore and interact with new technologies without the additional stress of knowing a grade depends on successful interaction/exploration, Ludlow assigns bonus points to activities incorporating new technologies for the first time and solicits feedback from the students on their perceptions of utility, feasibility, and acceptability of the technology. This allows students to become familiar with the technology before they use it to complete a graded assignment.

Instructors also should demonstrate understanding and compassion for students who are experiencing technology difficulties. Disregard for what, in some cases, may be true technology phobias or catastrophic computer failures only increases student anxiety and may contribute to course dropouts and negative attitudes (Collins et al., 2002). Recognizing that computer problems occur with everyone and acknowledging with a sense of humor the presence of the computer gremlin that has erased assignments can reduce tension and help learners persevere.

SUGGESTIONS FOR TROUBLESHOOTING PROBLEMS

Learning how to troubleshoot problems is no doubt one of most important skills online students and instructors must acquire. There are a number of ways that instructors can facilitate this process. Experienced online instructors recommend the following (Collins et al, 2002):

1. Structure activities to help students learn the skills they need to be successful. For example, begin with simple tasks that allow learners to become comfortable with technology in a non-threatening environment.
2. Provide step by step directions as indicated earlier for these activities so students can successfully post a message to the discussion board or participate in a chat.
3. Provide directions and examples for how to access course text, and provide online formats for assignments, projects etc., to reduce student confusion and errors.
4. Provide tips – complete work offline, then copy and paste….; save copies of work to hard drive; break large projects down into smaller tasks…; turn off call waiting function when online.
5. Model problem-solving and troubleshooting skills in class when problems occur during a webcast or in the online environment.
6. Post FAQs to the discussion board and encourage students to check there. Again, encourage students to use the "three before me" rule as well as assisting each other in problem solving.
7. Establish policies for problems such as server failure, inability to access content or a chat, computer crashes, or delayed access because of heavy traffic.

CONCLUSION

This chapter has described issues related to providing technical support for learners. Based on the professional literature and experienced online instructors, the author suggests that instructors consider the needs of the learner, the supports provided by the institution, and their role as instructor in ensuring sufficient technology assistance is available. Essential supports include a technology support infrastructure, "just-in-time" training opportunities, systematic orientation for students, clear course policies, and a calm and positive attitude on the part of the instructor.

REFERENCES

Bates, A.W. (2002). *Managing technological change.* San Francisco: Jossey-Bass.

Biner, P.M., Dean, R.S., Mellinger, A.E. (1994). Factors underlying distance student satisfaction with televised college-level courses. *The American Journal of Distance Education, 8*(1), 60-72.

Collins, B. C., & Baird, C. M. (this text). Online modules to prepare distance educators at the University of Kentucky

Collins, B.C., & Galyon-Keramidas, C. (this text). Planning and delivering online instruction.

Collins, B.C., Shuster, J.W., & Grisham-Brown, J. (1999). So you're a distance learner? Tips and suggestions for rural special education personnel involved in distance education. *Rural Special Education Quarterly, 18*(3/4), 66-71.

Collins, B.C., Shuster, J.W., Ludlow, B.L., & Duff, M. (2002). Planning and delivery of online coursework in special education. *Teacher Education and Special Education, 25,* 171-186.

Council for Higher Education Accreditation (2000). *The competency standards project: Another approach to accreditation review.*

Boulder, CO: National Center for Education Management Systems.

Foley, A. & Regan, B. (2002). Web design for accessibility: Policies and practice. *Educational Technology Review, 10*(1). Retrieved May 7, 2005 from http://www.aace.org/pubs/etr/issue2/foley.cfm

Gibb, G.S., & Egan, M.W. (2001). Guidelines for learner-centered evaluation of distance education programs. In B. Ludlow & F. Spooner (Eds.), *Distance education applications in teacher education in special education* (pp. 133-151). Arlington, VA: Teacher Education Division of the Council for Exceptional Children.

Hora, N., & Kling, R. (in press). Students' distress with a web-based distance education course: An ethnographic study of participants' experiences. *Information, Communication & Society.*

Jordan, L., Smith, S., Dillon, A.S., Algozzine, B., Beattie, J., Spooner, F., & Fisher, A.L. (2004). Improving content and technology skills in ADD/ADHD via a web-enhanced course. *Teacher Education and Special Education, 27*(3), 231-239.

Kearsley, G. (2000). *Online education: Learning and teaching in cyberspace.* Belmont, CA: Wadsworth.

Ko, S., & Rossen, S. (2004). *Teaching online: A practical guide.* Boston, CT: Houghton-Mifflin

Ludlow, B. L. (2005). *Web access letter for students.* (Available from Special Education Programs, West Virginia University, 606 Allen Hall, P.O. Box 6122, Morgantown, WV 26506-6122)

Ludlow, B. (2005). Technology-mediated distance education: Current practice and future trends. In D. Edyburn, K. Higgins, R. Boone (Eds.), *Handbook of Special Education Technology Research and Practice* (pp. 793-817). Whitefish Bay, WI: Knowledge by Design, Inc.

Ludlow, B.L., & Duff, M. (1998). *Distance education and tomorrow's schools.* Bloomington, IN: Phi Delta Kappa.

Ludlow, B.L., & Duff, M. (2001). Guidelines for selecting telecommunications technologies for distance education. In B. Ludlow & F. Spooner (Eds.), *Distance education applications in teacher education in special education* (pp. 17-54). Arlington, VA: Teacher Education Division of the Council for Exceptional Children.

Ludlow, B.L., & Duff, M. (2002). Live broadcasting online: Interactive training for rural special educators. *Rural Special Education Quarterly, 21*(4), 26-30.

Menlove, R., & Lignugaris/Kraft, B. (2004). Preparing rural distance education preservice special educators to succeed. *Rural Special Education Quarterly, 23*(2), 18-26.

Owsten, R.D. (1997). The world wide web: A technology to enhance teaching and learning? *Educational Researcher, 26*(2), 27-33.

Pennsylvania State University. (1998). *Innovations in distance education: An emerging set of guiding principles and practices for the design and development of distance education*. State College, PA: Author.

Pindiprolu, S.S., Peck Peterson, S.M., Rule, S., & Lignugaris/Kraft, B. (2003). Using web-mediated experiential case-based instruction to teach functional behavioral assessment skills. *Teacher Education and Special Education, 26*(1), 1-16.

University of Illinois (n.d.). What makes a successful online learner? http://www.ion.uillinois.edu/resources/tutorials/pedagogy/StudentProfile.asp [retrieved April 2, 2005]

Varvel, V. (2001). Facilitating every student in an online course. http://www.ion.uillinois.edu/resources/pointersclickers/2001_03/index.asp [retrieved April 2, 2005]

Whitworth, J.M. (1999). Looking at distance learning through both ends of the camera. *The American Journal of Distance Education, 13*(2), 64-73.

Part 2

PROGRAM DESCRIPTIONS

Chapter 4

Online Delivery of Programs in Low Incidence Disabilities and Special Education Administration to Meet Statewide and National Needs

Harvey A. Rude
University of Northern Colorado
Kay Alicyn Ferrell
National Center on Low Incidence Disabilities

Title of Program/Sponsor

Master of Arts degree and teacher licensure in low-incidence disability areas (Deaf/Hard of Hearing, Visual Impairment, Severe Disabilities) and post-Masters and administrator licensure (Director of Special Education) at the University of Northern Colorado

Brief Description of the Program

Overview of program. The Master of Arts degree program in Special Education at the University of Northern Colorado (UNC) provides a variety of emphasis areas that lead to teacher licensure/endorsement in specialized areas of teacher preparation. Among the eight emphasis areas in this degree are three low-incidence categories of Deafness/Hard of Hearing, Visual Impairment, and Severe Disabilities. A post-Masters endorsement program is available for individuals seeking an administrator's license with a Director of Special Education endorsement. All coursework for each of these programs is provided through online instructional formats designed to meet the needs of participating graduate students who live in rural and remote locations of Colorado, the Western region of the United States, and other national or international locations where such programs are not offered by local universities.

Areas of Specialization

Deafness/Hard of Hearing
Visual Impairment
Severe Disabilities.

Application

Master's degree with teacher licensure/endorsement in low incidence disabilities

Post-Masters endorsement program for Director of Special Education

Program Level

Pre-service personnel preparation

Target Audience
Graduate students who live in rural and remote locations

Online Component
All courses offered entirely online

TECHNOLOGY APPLICATIONS USED FOR ONLINE INSTRUCTION

Course Management System/Software and Hardware

The programs utilize the Blackboard course management platform. Blackboard is a widely adopted platform that supports both synchronous and asynchronous interactions among participants through ongoing chat rooms and discussion boards.

Asynchronous Online Formats
(Special Education Administration Program)

Presentation of content. In the special education administration program, the provision of supporting resources and media is accomplished through a combination of annotated PowerPoint presentations, full text documents, audio and video clips, and web links. A structured instructional approach is provided through sequential units of instruction that include all lecture notes, application activities, course readings, chat rooms, online conferencing, email, listservs, message boards, and threaded discussion boards. The availability of an electronic grade book provides students with an ongoing opportunity to monitor the status of their assignments and major course projects.

Interactions with learners. The keys to successful online course experiences are (a) frequent ongoing interactions between professor and students, (b) frequent ongoing interactions among all course participants, and (c) timely feedback on all assignments and inquiries posed by online learners. In addition to regular online office hours, submission of all assignments and course questions is encouraged through ongoing email communication with a 24-48 hour response.

Asynchronous Online Formats
(Low Incidence Disabilities Program)

Presentation of content. The programs that prepare teachers of learners with low-incidence disabilities employ a somewhat different approach, given that about 10% of the enrollment is composed of individuals with disabilities. The courses have been created from a philosophy of universal design, making sure that course content is accessible to every student, regardless of disability, utilizing the principle of least restrictive format. For example, PowerPoint presentations are graphic images, unintelligible to a screen reader. National Center on Low-Incidence Disabilities (NCLID) staff, therefore, convert instructor-created PowerPoint presentations to HTML and XML formats and try to eliminate instructor reliance on photos or other graphic images. Other parts of Blackboard do not meet accessibility standards for persons with disabilities; when that occurs (as it does with Blackboard's chat rooms), NCLID has established accessible chat rooms on another server. Some faculty members also utilize audio feeds accompanied by written transcripts, and many create their own videos or CDs, purchased by students through the university bookstore. (Video streaming is not used as this time, because many students are still using dialup rather than high speed internet connections.) In cases where only a commercial videotape meets the instructional need, several copies are purchased by the program, circulated to students, and returned to the program before grades are due.

Interactions with learners. The low-incidence programs also employ a constructivist approach to learning. Rather than place artifacts left over from face-to-face classes online (such as lecture notes, transparencies, or handouts), faculty have tried to focus on the tools needed to do the job. Activities and assignments simulate real life experiences as much as possible. While students still have significant reading assignments, instructors have scaffolded learning through guided experiences that lead up to outcomes applicable to the everyday demands of an itinerant teacher. These guided experiences are addressed primarily by case studies, but they are marked by what we call "social engagement." Students are encouraged to work online with each other, forming their own workgroups, and faculty provide a constant flow of information and questions through the use of email listservs. At times, expert teachers are invited into the discussion, to provide a new perspective and a fresh dose of reality.

Many of the low-incidence courses require students to complete activities or assignments weekly. Faculty provide feedback on these activities either by listserv, with generalized statements of interest to all students; by using the "track changes" function in Microsoft Word for individual assignments; or by converting the student's document to PDF, where notes, comments, highlighting and editing can be added using the complete version of Adobe Acrobat. This last method is particularly useful, because Adobe Acrobat Reader is a free download, and all comments can be viewed at one time. However, PDF files still create problems for screen readers, so the method is not used with blind and visually impaired students.

FACTORS INFLUENCING DECISION TO USE ONLINE INSTRUCTION

NEEDS ADDRESSED BY ONLINE INSTRUCTION

The low-incidence master's degree and teacher/administrator licensing programs delivered from Northern Colorado were created as online options in response to critical shortages of educators in rural and remote areas of Colorado and the Western regions of the United States.

REASONS FOR SELECTION OF SPECIFIC ONLINE FORMATS

Prior to 1998, all courses required of educators planning to become qualified as a teacher of students with low-incidence disabilities (e.g., learners who are deaf/hard of hearing, visually impaired, or who have severe disabilities) or licensed Directors of Special Education were delivered through on-campus formats that required prospective graduate students to be in residence at the Greeley campus. The option of providing off-campus cohort-driven programs, or even through compressed video or video conferencing methodologies, was not a viable alternative due to the vast geographical distribution of potential students. Consequently, students were expected to become residential, campus-based students for several consecutive semesters or participate through a "summers only" commitment. Low numbers of students committed to these delivery options (prior to online course delivery the number of participating students in each of the four programs was between 15-25 students) resulted in speculation concerning the viability of continuing these programs at relatively high-cost that were undersubscribed in comparison to high-incidence special education and regular education teacher education programs. An additional concern was that students in the summers only program were taking too long to complete a degree, often spanning the maximum time for a degree (5 years) and sometimes failing comprehensive examinations (which we believe was due to a lack of continuity throughout the year).

OUTCOMES AND IMPLICATIONS OF ONLINE INSTRUCTION

Impact of Online Instruction on This Program

The commitment to convert existing graduate degree/licensure programs to online formats was accomplished over a period of several years. The number of active students in each of these programs increased dramatically as a result. In fall semester of 2004, the numbers of active graduate students was as follows: (a) Deaf/Hard of Hearing – 56, (b) Visually Impaired – 70, (c) Severe Disabilities – 47, and (d) Director of Special Education – 104. The low-incidence disability programs are the only programs of this nature in the state, and they serve as a regional resource of teacher preparation for many other states in the western United States. Prior to converting to online delivery, there were two in-state programs preparing Directors of Special Education; at the present time, there is one surviving program in the state. A significant factor in the enrollment growth and viability of each program is the provision of all classes through the online instructional approach. The practicum and internship experiences required of each program graduate is accomplished through qualified adjunct faculty supervisors in local school districts or community agencies where the students reside. An online course supplement is available to support each graduate student completing the required field experiences. An additional incentive for students to participate in these programs is the reduced tuition structure provided to students enrolled in the low-incidence programs as a participating Western Regional Graduate Program approved through the Western Interstate Consortium for Higher Education (WICHE).

Opportunities/Advantages/Benefits of Online Instruction

The development and ongoing support for online programs has been achieved in the face of resistance from instructors who are not adept or comfortable in the online instructional paradigm. The most common concerns expressed by faculty who are uninitiated in the process of online learning are: (a) the lack of transfer between the online classes and the demands of the job, (b) failure to accommodate learning processes thereby creating disruptions in learning, and (c) the attrition of online students who do not complete individual classes and the programs they support. In fact, the low-incidence programs seem to be very attuned to the demands of the real world, address a variety of learning styles, and have experienced less attrition than they did during the on-campus years. However, to address these concerns, the UNC programs acknowledge and support the need for three different types of online learning including receptive, directive, and guided discovery. Receptive learning is used to inform course goals, and provides significant information acquisition with limited practice opportunities. Directive learning is designed to strengthen performance on procedural goals through frequent responses from learners encouraged by timely feedback. Guided discovery seeks to construct new knowledge that informs job embedded competencies by solving real problems supported by appropriate resources. The design of each online course blends these three major purposes of electronic learning in a manner that encourages performance improvement.

Problems/Disadvantages/Limitations on Online Instruction

These approaches are not implemented without considerable faculty investment. A common misconception among educators is that online instruction is a ruse constructed to give faculty more time away from campus or more time to pursue other interests. However, our experience is that the courses require considerable investment of time and resources up front, in order to design the courses to accomplish objectives and address standards. They also require ongoing and continuous engagement by faculty to sustain social interaction and learning. Faculty have developed various methods for handling the amount of work involved just in email traffic, from holding virtual office hours to respond to questions, to developing an asynchronous discussion board for questions about the class, to using listservs for group comments, to using rules to sort email into separate folders for each course taught. However, while online courses offer flexibility in terms of faculty time, there is no doubt that the amount of time involved in instruction has increased geometrically for faculty – actually extending the work day/work week, instead of shortening it.

Suggestions or Recommendations for Others Considering Online Instruction

A useful feature of all online classes and experiences at UNC is the considerations for applying universal design and the redundancy principle. The universal design principle indicates that it is not necessary to use all available "bells and whistles" simply to be cute in your course design. All online classes are created using the least restrictive environment that will facilitate information acquisition and application by participating students who may themselves have a low-incidence disability, such as deafness or a visual impairment. This is further supported by the redundancy principle or the commitment to provide multiple opportunities to practice key concepts that has consistently been proven to demonstrate higher levels of learner mastery. As learners in an online environment are provided with significant opportunities to practice the intended outcomes of a course, the speed and accuracy of response both increase.

Another guiding ideal of online class instruction is the focus on teaching learners to explain examples of key concepts in their own words and experiences. The principle of self-explanations has been demonstrated to enhance long-term memory, reinforcement of preferred learning processes, and the degree of knowledge, skill application, and mastery. The use of structured collaborative techniques is valuable in promoting the necessary components of skill practice that lead to the application of skills in the applied setting on the job. Some examples of structured collaboration in the online environment include jigsaw, structured controversy, problem-based learning, and peer tutoring. The outcomes from these techniques have included (a) tangible products, such as program evaluation models and/or logic models of planned change; (b) group processes, such as inter-agency strategic action plans and/or collaborative case studies; and (c) process outcomes, including strategic thinking options and/or the identification of adaptive challenges and subsequent intervention activities. Structured group assignments have proven successful when based on appropriately designed cases supported by appropriate resources. The online instructional tools that are most effective in support of the collaborative outcomes include structured study assignments, threaded discussion boards, email interactions to follow-up on specific questions as well as evaluate projects, and chat rooms or other conferencing techniques to provide systematic feedback.

An interesting finding from online course instruction is the extent of outcomes from learners in primarily learner-controlled versus program-controlled environments. In general, the participants with less knowledge of or prior experiences with the course content benefit to a greater extent from program-controlled approaches to online learning. The learner-controlled environment allows participants to bypass certain aspects of the online course material to proceed to areas of greatest interest or impact. For example, a participant with prior experiences in coordinating special education services or functioning as a department chair will make the connections among various program content and experiences with greater facility than someone without these related experiences. Another way of explaining this phenomenon is that novices spend an extensive amount of time exploring course materials, whereas experts (i.e., those with more sophisticated understanding of program content and contexts) invest significantly greater amounts of time in the processes of analysis, planning, and implementation of intended outcomes.

Learner controlled environments are not synonymous with expecting or necessarily allowing students to skip information or "surf" through course materials in a random fashion. The online courses are built with requirements redundantly displayed everywhere with multiple opportunities to reinforce concepts and ensure mastery of outcomes. The learner-controlled environment encourages a great degree of freedom in approaching the available content. This provides considerable room for controversy, and this controversy breeds discussion and problem solving.

The key online principles to be observed in differentiating online course experiences include personalization, redundancy, modality, coherence, multimedia, and contiguity. These principles support distinctive types of online learning that include the show and tell receptive, tell and do directive, and problem-solving guided discovery approaches for optimal outcomes from diverse learners. The most significant instructional question that is addressed continually by these online programs is how to achieve the intended outcomes that are at least equivalent to face-to-face instructional delivery from the online classes and experiences. With few exceptions, we believe our online courses produce even greater outcomes, because of the capacity of the technologies to create cohesive and committed communities of practice across vast geographic areas. Those connections persist long after the course concludes – with and without the instructors.

LIST OF RESOURCES

Books and Articles

Clark, R. C., & Mayer, R. E. (2003). *E-Learning and the science of instruction: Proven guidelines for consumers and designers of multimedia learning.* San Francisco, CA: Pfeiffer.

Mayer, R. E. (2001). *Multimedia learning.* New York: Cambridge University Press.

Reeves, B., & Nass, C. (1996). *The media equation: How people treat computers, television, and new media like real people and places.* New York: Cambridge University Press.

Web Sites

Access E-Learning, http://www.accesselearning.net/

Georgia Tech Research on Accessible Distance Education, http://www.catea.org/grade/

Phaedrus Academy: Online Course for Teachers Learning To Teach Online. http://vision.unco.edu/edse/Phaedrus/announce.html

Project EQUIP (Educational Quality through Universal Instruction Principles), Faculty Resources, http://www.unco.edu/equip/Resources/resources_faculty/default.asp

Project EQUIP (Educational Quality through Universal Instruction Principles), Universal Design Website Links, http://www.unco.edu/equip/UD%20References/default.asp

CONTACT INFORMATION

Harvey A. Rude, Ed.D.
Professor and Director
School of Special Education
University of Northern Colorado
Greeley, CO 80639
970-351-1659
Harvey.Rude@unco.edu
http://www.unco.edu/coe/debe/index.asp

Kay Alicyn Ferrell, Ph.D.
Professor and Executive Director
National Center on Low-Incidence Disabilities at the University of Northern Colorado
Greeley, CO 80639
970-351-1653
Kay.Ferrell@unco.edu
http://www.nclid.unco.edu

Chapter 5

Online Master's Degree Program in Special Education at East Carolina University

Sue Byrd Steinweg
Sandra Hopfengardner Warren
East Carolina University

Title of Program/Sponsor

MAEd in Special Education
Department of Curriculum and Instruction, College of Education at East Carolina University

Brief Description of Program

Overview of Program. The Master of Arts in Education – Special Education (MAEd-SPED) program at East Carolina University (ECU) is a part-time graduate degree program for individuals holding (at least) an initial special education license. The program has served as North Carolina's only totally online MAEd-SPED program since Fall, 2002. Courses within the program address the standards of accrediting agencies and result in attainment of advanced teaching competencies. The program presents coursework available through the Internet so students may access course materials at their convenience. Practicum experiences are integrated within classes enabling students to use their own special education classroom as the practicum site. The 39-semester hour program includes 13 courses that are offered using a cohort model enabling students to complete the program in four semesters and two summer sessions. A new cohort of 20 students is accepted each fall and spring semester. Students register for two classes each semester and each of the two summer sessions. They are required to complete four MAEd core courses, four professional studies courses in Special Education, and five disability-specific courses, including specialty area research. Since the vast majority of students are from North Carolina, the MAEd-SPED focuses on the North Carolina curricula (i.e., Standard Course of Study). However, it is designed to be flexible enough to enable out-of-state students to address their respective state's curricula. Students also may elect to pursue the 12-semester hour online ECU assistive technology certificate.

Area(s) of Specialization

Learning Disabilities
Mental Retardation
Behavioral and Emotional Disorders
Low Incidence Disabilities/Severe and Profound Disabilities

Application

Master's degree program

Program Level

Graduate, advanced special education teacher licensure

Target Audience

Educators holding clear special education teacher licensure

Online Component

All courses are offered entirely online; learners are not required to come to campus at any time

TECHNOLOGY APPLICATIONS USED FOR ONLINE INSTRUCTION

Course Management System/Software and Hardware

Blackboard is the course management system used for courses and is available to all faculty and students at ECU. Courses are specifically designed for students who do not have high-speed Internet access. Since the university contracts with faculty members for development of each new online course, course frameworks are available for sharing across faculty. Course materials are created in a variety of formats using Macromedia Dreamweaver, PowerPoint, and Microsoft Word. Documents are presented as PDF files or in Rich Text Format within Blackboard. Course materials that involve video or formats that would require high-bandwidth to access are burned to CDs and mailed to students. When coursework involves unique equipment or materials, these items are mailed to the student for use during the semester. For example, students in the assistive technology course receive an AT Tool Kit containing over $2,000.00 worth of hardware and software.

Asynchronous Online Formats

Presentation of content. Content is presented in varying formats across courses, including text, images, audio and video clips, case studies, links to Web sites, and electronic course packs. Students in the assessment course receive a CD developed by an ECU faculty member demonstrating test components and appropriate assessment administration. Course assignments emphasize application of content to the student's work setting. Students are required to purchase and read print resources, including textbooks or other professional books, and to access professional articles through the ECU Virtual Library electronic resources.

Interactions with learners. Interactions with learners and among learners are fostered through the Blackboard components of discussion boards, virtual classroom, group pages, and digital drop box. Students are required to participate in discussion board asynchronous discussions on topics posted by instructors or peers. Discussion board postings include presenting information on readings or online resources, analyzing curriculum materials or websites, brainstorming solutions to challenging situations, discussing case studies, summarizing professional articles, sharing professional experiences, and discussing key issues. Students are frequently divided into groups to engage in projects or provide peer review. Some courses include position papers, video taped demonstration of skills, group projects, and electronic poster presentations. Assessment in courses is primarily through products that demonstrate application of course content. Feedback is provided through email, electronic hand-written comments, and phone conversations. Students work with instructors to develop, implement, and present thesis or action research projects. They also develop electronic portfolios to demonstrate growth through the program. Portfolio frameworks reflect the advanced competencies of accrediting agencies and professional organizations.

Synchronous Online Formats

Presentation of content. Some courses include synchronous communication through the virtual classroom to provide clarification and discussion.

Interaction with learners. Interactions among learners may occur in the Blackboard virtual classroom for group projects or activities. Some instructors use the virtual classroom function in Blackboard to conduct online office hours. Students present findings from thesis or action research to faculty and peers.

FACTORS INFLUENCING DECISION TO USE ONLINE INSTRUCTION

Needs Addressed by Online Instruction

Rural areas in North Carolina and surrounding states have had a shortage of highly qualified special education personnel that has been amplified by a high attrition rate. ECU's previous MAEd-SPED program served small numbers because of the long travel time involved for many teachers who wanted to pursue a Master's degree. In 2001, ECU obtained state approval to convert the program to a totally online format to facilitate training of an increased number of master teachers throughout the state.

Reasons for Selection of Specific Online Formats

ECU had selected the Blackboard course management system as a way to provide a package of features that allow for the presentation of course content and activities. The standard format for courses is helpful for students as they do not have to relearn course structure with each course. The primarily asynchronous format allows the program to serve teachers in rural areas who may not have high-speed Internet access.

OUTCOMES AND IMPLICATIONS OF ONLINE INSTRUCTION

Impact of Online Instruction on This Program

The online MAEd program has provided online coursework for more than 100 students since 2002. The first cohort of 18 part-time MAEd-SPED students graduated in December, 2004 and 2 full-time students graduated in May, 2004. The totally online format has greatly increased interest in the program for individuals residing further from the university and has resulted in a dramatic increase in enrollment. The growth in number of students also has led to the addition of faculty and resources for the program.

Opportunities/Advantages/Benefits of Online Instruction

Some of the benefits of online instruction include (a) increased number of master teachers for the state, especially in rural areas; (b) higher enrollment resulting in additional resources for the program; (c) increased responsiveness to students through electronic formats; (d) increased student involvement in the learning process; and (e) improved instruction through multiple formats and options.

Problems/Disadvantages/Limitation of Online Instruction

In many respects, presenting coursework online is more complicated than traditional face-to-face delivery as it requires more time for faculty in developing of coursework, responding to students, and providing feedback. Faculty also must invest time in increasing their own technology skills. Insufficient rapid response to technical problems will result in frustration for students and faculty. Careful consideration by instructors is critical for developing learning communities without the benefit of face-to-face interactions.

Suggestions or Recommendations for Others Considering Online Instruction

Developing a totally online program requires a level of commitment from the university and program administration to provide resources and time necessary to purchase technology, provide technical support for faculty and students, provide training for faculty, support time for course development, and facilitate access for students with disabilities. Consideration should be given for how online instruction will be evaluated for tenure seeking faculty and what incentives will be provided for faculty for developing online courses. These incentives may include graduate student support, improved technology, training opportunities, and stipends or release for course development. Careful consideration must be given to the materials used in online classes as the copyright laws governing resources that can be used online vary from those governing face-to-face classes. Programs must consider the audience and their technology resources. If the program is focused on rural areas with low bandwidth access it will be necessary to develop courses that can be accessed by these individuals. Courses should be designed for use with available hardware and free or inexpensive software to minimize difficulties resulting from limited technology resources of many students and the potential lack of local technical support. This may include developing creative ways to provide the content and resources needed by students. Consideration should be given to how technical support will be provided for online students. Student attrition may be a problem with online courses because of technical difficulties, feelings of isolation, and student responsibility. These difficulties can be addressed by developing a program where expectations for courses are clearly stated, a high level of student interaction is fostered, online technology support is provided, and faculty provide timely feedback and encouragement.

LIST OF RESOURCES

Books

Dabbagh, N., & Bannan-Ritland, B. (2005) *Online learning: Concepts, strategies, and application.* Upper Saddle River, NJ: Pearson.

Gunawardena, C., & McIsaac, M. (2002). Distance education. In D.H. Jonassen (ed.), *Handbook of research on educational communications and technology* (pp. 355-395). Mahwah, NJ: Erlbaum.

Hanna, D., Glowacki-Dudka, M., & Conceicao-Runlee, S. (2000). *147 practical tips for teaching online groups: Essentials of web-based education.* Madison, WI: Atwood.

Moore, G., Winograd, K., & Lange, D. (2001). *You can teach online: Building a creative learning environment.* New York: McGraw Hill.

Moore, M., & Kearsley, G. (2005). *Distance education: A systems view.* Belmont, CA: Thomson Wadsworth.

Palloff, R., & Pratt, K. (2003). *The virtual student: A profile and guide to working with online learners.* San Francisco: Jossey-Bass.

Palloff, R., & Pratt, K. (2005). *Collaborating online: Learning together in community.* San Francisco: Jossey-Bass.

Web Sites

http://www.blackboard.com

CONTACT INFORMATION

Sandra Hopfengardner Warren, Ph.D.
Special Education Program Area Coordinator
Department of Curriculum and Instruction
125 Speight
East Carolina University
Greenville, NC 27858
252-328-2699
warrens@mail.ecu.edu
http://www.coe.ecu.edu/sped/spedhome.htm

Chapter 6

Indiana University's Collaborative Teacher Education Program

Dennis Knapczyk
Professor of Curriculum and Instruction
Indiana University

Title of Program/Sponsor

Collaborative Teacher Education Program (CTEP)
Indiana University

Brief Description of Program

Overview of program. Indiana University's Collaborative Teacher Education Program (CTEP) provides prospective teachers with the knowledge and skills to engage in school-based collaborative planning that assists children and youth with disabilities in participating successfully in general education settings. CTEP is a joint program across four campuses of Indiana University with a geographic region encompassing most of central and southern .Indiana. The program makes certification in mild (high incidence) disabilities and emotional/behavior disorders accessible to teachers through distance education. Our primary target audience is teachers in rural communities who are teaching on limited/emergency licenses or who are pursuing new careers in special education, and we have over 200 teachers in our program. CTEP offers 30 credit hrs of graduate level coursework which are scheduled over a 3-yr time span. During the 20 years CTEP has been in operation, we have used various distance education technologies including speaker phones, audiographics and videoconferencing. Now, all of our courses have major online components, and we have five courses that are completely online with others to follow. The hallmarks of CTEP are that each course (a) uses a performance-based structure to address multiple state and national standards and (b) links assignments and activities directly to on-the-job field applications.

Area(s) of Specialization

Mild (High Incidence) Disabilities
Emotional/Behavior Disorders

Application

Graduate level certification for teachers

Program Level

Pre-service personnel preparation

Target Audience

Teachers in rural communities who are teaching on limited/emergency licenses or who are pursuing new careers in special education

Online Component

Various distance education technologies including speaker phones, audiographics and videoconferencing, with major online components and five courses that are completely online with others to follow

TECHNOLOGY APPLICATIONS USED FOR ONLINE INSTRUCTION

Course Management System/Software and Hardware

We use Indiana University's Oncourse software program for online course instruction, practicum supervision, general advising and other program-related tasks. Oncourse is a web-based instructional support program that was specifically designed to assist faculty and students with a variety of course-related functions. For example, in Oncourse there are folders and work areas where we can display a course syllabus, make announcements, send handouts and post grades; where teachers can show their pictures, make personal profiles, send one another email and put assignments in drop boxes; and where class groups can hold discussions, have chats and collaborate in various ways on activities and projects. We conceptualize our online courses as having three major components: (a) course content and readings; (b) activities and practice exercises; and (c) projects and field work applications.

Asynchronous Online Formats

Presentation of content. This component refers to the theoretical concepts, paradigms, models and other basic information normally covered in texts and course readings that provide the foundation for learning about teaching methods. We often use two types of online activities to cover course content: (a) weekly large group discussions (Mondays through Fridays) and (b) weekly small team reflections (Saturdays and Sundays). For example, we might have a teacher serve as what we call a Starter for the week. On Monday morning, this person posts a summary of the assigned readings and three questions designed to stimulate the large group discussion for the week. We post a model for Starters to follow so their postings meet our expectations for such things as the amount of detail to use in summarizing readings and the types of questions to ask. The other students act as Participants in the discussion by responding to each of the Starter's questions and making 1-2 additional postings commenting on their classmates' works. On Friday afternoon, another teacher acts as a Wrapper and summarizes the week's discussion, such as highlighting the group's interaction, noting key concepts discussed, pinpointing areas of controversy, and identifying important applications of concepts. On weekends, we often have teachers work in small teams where they might post a reflective summary of what they learned during the group discussion and comment on their classmates' reflections.

Practice exercises are the opportunities we structure for teachers to try out concepts before actually using them in their teaching situations. Our goal with these activities is to broaden their understanding of course material by having them explain, individually or collectively, how they would apply concepts, principles and guidelines in on-the-job situations. In our online format, we often have teachers work in teams of 3 to 5. We might provide a case study and have each team work on questions or tasks in which they engage in collaborative problem solving or planning. Team members might make individual postings that include examples of concepts and methods, teaching applications, sample materials and case reviews. Team members also would read and reply to their teammates' postings with suggestions or critiques. Often, we would have teams integrate their discussions into a final product, such as reporting on an assessment profile, proposing a teaching lesson or suggesting applications of instructional methods or materials

The third component of our online courses is fieldwork applications where teachers use course concepts in their own teaching situations. Our goal here is to have teachers demonstrate their ability to integrate and adapt best practices in real life circumstances. To accomplish this goal, we give project-based assignments that involve multi-step tasks that build toward field application of several course concepts. The projects reflect state licensing standards for the topic of the course so teachers also can use their finished products to demonstrate competence in these areas. We provide a structure for the projects that includes a description of project outcomes, the components of the finished product, the tasks teachers need to complete, and a grading rubric. We also provide a model of the product or samples of certain components. For example, in our course on Assessment and Instruction, we assign two projects to cover field work applications: (a) Project 1 – Conducting a curriculum-based assessment on a student with a disability and using the results to prepare an Individualized Education Plan (IEP), and (b) Project 2 – Developing a thematic unit for an academic area that incorporates differentiated instruction and principles of diversity. In our course on Teaching Social Skills, teachers do one project in which they assess the social skills of a group of students and plan an instructional unit based on the results, and another project in which they develop a school wide strategy for improving social skills in their building. We often provide teachers with online discussion folders where they can post their work, discuss directions and activities, solicit suggestions, and collaborate on a final product.

We require teachers to do at least one practicum in order to obtain a teaching license or master's degree, and they do it in their current teaching assignment. We link the practicum to a content course they are taking so they learn and apply best practices at the same time. The practicum we offer each semester is topical and changes from semester to semester, depending on the courses we schedule. As teachers go through a content area course, such as Classroom Management or Assessment and Instruction, those enrolled in a practicum would do a semester-long project that concentrates on field-based applications in that area. For example, when we offer the Management of Behavior Disorders course, we focus the practicum that semester on activities associated with behavior management. We would assign teachers a semester-long project in which they plan, carry out and evaluate an intervention for a student in their classroom who displays severe and longstanding emotional/behavior disorders. We give them written directions that divide the project into four parts: (a) describing the student they selected, the problems he/she displays and the circumstances in which the problems occur; (b) establishing a priority problem area for intervention and preparing an intervention plan; (c) implementing and evaluating the intervention plan; and (d) reflecting on the intervention process. In the directions, these four parts are further divided into 28 steps or tasks. Along with the directions, teachers are given a grading rubric showing that their final grade would be determined by such factors as the content, organization, writing style, creativity and timeliness of their work. When we schedule the Teaching Social Skills or the Transitions Across the Lifespan course, we have comparable semester-long projects for teachers to work if they choose to do their practicum. Teachers can do a practicum anytime in the program so they can select the particular topic on which to focus their work.

Teachers work on their practicum in a mentorship arrangement. We use as mentors teachers who are enrolled in the content course that is linked to the practicum. For example, if teachers are focusing their practicum in the area of classroom management, their mentors would be other teachers who are enrolled in the Classroom Management course offered that semester. A portion of the course grade for Classroom Management would be based on the work mentors do in mentoring a practicum student so the course and the practicum are designed to support one another. That is, the practicum students' role is to complete a field-based project, such as on classroom management, and the mentors' role would be to assist the teachers in their work. Each teacher doing a practicum has 1-3 peer mentors, depending on the enrollment in the content course.

The online environment serves as a critical tool in supervising practicum projects and facilitating interactions among teachers and their mentors. In their practicum, teachers maintain an online weekly journal in which they give a running account of their work on each step of the project. They post each journal entry for their mentor(s) and instructor to review and comment on. We also have them to engage in an online dialogue with their mentor(s) and instructor by asking questions, seeking their input, responding to suggestions and feedback, and interacting in other ways. These at least weekly interactions are in addition to the regularly scheduled journal entries. The mentors, on the other hand, are instructed to serve as online consultants to the practicum students by doing such things as keeping informed on the teacher's progress, asking questions to clarify decisions and activities, offering suggestions and critical feedback, and, in other ways, acting as a professional resource. We emphasize the importance of showing timeliness, responsiveness and initiative in interactions with the teacher. We give mentors a schedule for their postings which is to respond within three days to each of the practicum student's journal entries.

At the start of the practicum, each teacher is designated the team leader for his/her project and the teacher's mentor(s) and instructor are the other team members. Each team has its own online discussion folder in Oncourse, and members use the folder to hold a semester-long dialogue regarding practicum activities and other pertinent topics. For example, team members use the folder to describe their personal backgrounds, share professional experiences, provide progress reports, ask for clarification of activities, offer suggestions, give critical feedback and socialize. A teacher's journal postings and the interactions among the teacher, mentor(s) and instructor are organized into one semester-long discussion thread. The Oncourse Discussion Folder also permits teachers to post attachments, such as a graph of their student's baseline performance or a table showing a lesson plan.

Interactions with learners. In our online courses, we see the instructor's primary role as the facilitator or moderator of learning rather than as the primary source of information. As facilitators, we begin the semester by encouraging cooperative behavior, suggesting online conversational techniques, modeling posting and interaction formats, and providing other guidance in meeting course expectations. After a couple of weeks, we step back from group discussions and let teachers take charge of their interactions in response to classmates' postings or to assignment directions. To avoid providing knowledge directly and thus promoting instructor-centered learning, we usually ask probing questions to invite teacher responses and critical thinking. Our experience shows that teachers quickly take over responsibility for discussing ideas and begin asking and answering questions of each other, but the instructor still closely monitors student interaction throughout a course to clarify misunderstandings and direct discussions in productive ways.

The instructor's role in the online practicum is to play a facilitative role in activities and act as another mentor to teachers. Instructors usually begin with ice breaker activities that encourage socialization among team members, give each member a personal identity, and build comfort and trust among participants. They also model use of personal greetings, conversational language and other online strategies for enhancing communication. During the rest of the semester, they review journal postings; participate in team discussions; and offer suggestions, encouragement and constructive feedback. In practice, the nature and extent of this role varies considerably from team to team. For example, some teachers might be unclear of the differing roles of practicum students and mentors or of specific task expectations. Instructors would post information addressing these concerns on the Oncourse webpage or as an email message to team members. Similarly, a practicum student might be unclear about the required length and detail of journal postings, the timeline for project tasks, the level and the type of interaction needed to complete activities and other course-related demands. For example, it is sometimes difficult for teachers to gauge the amount of background information to give in a posting so other team members will be able to understand the context for decisions or proposed activities, and the instructor would have to step in and explain what is needed. Therefore, instructors sometimes have to be more directed in their approach to overseeing activities to address problems like these.

FACTORS INFLUENCING OUR DECISION TO USE ONLINE INSTRUCTION

Needs Addressed by Online Instruction

In the last couple of years, we have been transforming our program to an online format not only to increase its accessibility to teachers but to also to broaden instructional options in coursework and field experiences.

Reasons for Selection of Specific Online Formats

We have been transforming CTEP coursework from a videoconferencing to an online format, and there were several compelling reasons for starting the changeover, such to improve program accessibility, add flexibility to instructional activities, save on program costs and increase the pool of available instructors.

OUTCOMES AND IMPLICATIONS OF ONLINE INSTRUCTION

Opportunities/Advantages/Benefits of Online Instruction

We are discovering substantial benefits from using online instruction compared to conventional on-campus and videoconferencing approaches. For example, we found that in class discussions, the asynchronous format of online learning gives teachers more time to prepare a thoughtful response. Previously, topics and questions in discussions were often not well explored because teachers did not have sufficient time to frame a full response and, as a consequence, their statements were often spontaneous, shallow and incomplete. By comparison, in an online discussion, teachers have time to read over questions, think through an answer, prepare a response, and review and edit it before posting it to the discussion network. Furthermore, an online discussion can proceed for days or longer so teachers have the opportunity to read the responses of their classmates, ask for clarification of ideas, consider differing viewpoints, and re-formulate their own responses if they wish.

We also found that using an online format for practicum supervision and mentoring has some unique advantages over traditional ways of offering field experiences. In a practicum for limited licensed teachers, a natural context for their work is learning to use and apply best practices in their current teaching assignment. But in rural communities, teachers doing their practicum are usually scattered across a large geographic region. An online format allows teachers to have mentors and a supervisor who are geographically distant from them, giving them access to professional expertise and assistance that might not otherwise be available in the school or community where field experiences take place. Trainees and mentors can interact frequently and at more convenient times online instead of trying to fit meetings into busy work schedules. The asynchronous aspects of online mentoring also facilitate more thoughtful, task-oriented interaction than face-to-face discussion that can sometimes be too fragmented and incomplete to address complex problems. Finally, an online format provides greater privacy and anonymity than face-to-face communication so trainees are more apt to address sensitive and personal concerns.

SUGGESTIONS OR RECOMMENDATIONS FOR OTHERS CONSIDERING ONLINE INSTRUCTION

In summary, we have found that online learning is a valuable tool for preparing teachers on limited licenses because it makes coursework and practicum experiences more accessible to them. Moreover, this approach allows them to share experiences and concerns, engage in problem solving, and develop and test out new ideas. Online learning can provide a collaborative tool to promote cooperative learning, problem solving and professional support.

LIST OF RESOURCES

BOOKS AND ARTICLES

Knapczyk, D. (In press). *Professional development in a web conferencing environment.* Technical Horizons in Education Journal.

Knapczyk, D., Chapman, C., Rodes, P., & Chung, H. (2001). Teacher preparation in rural communities through distance education. *Teacher Education and Special Education, 24,* 402-407.

Knapczyk, D., Rodes, P. Chapman, C., & Chung, H. (2001). Collaborative teacher education in off-campus rural communities. *Rural Special Education Quarterly.*

CONTACT INFORMATION

Dennis Knapczyk, PhD.
Professor of Education
Department of Curriculum and Instruction, Special Education

Program Area
Education 3238
201 Rose Avenue
Bloomington, Indiana 47405
812 856-8148
Knapczyk@indiana.edu
CTEP Website: www.indiana.edu/~ctep

Chapter 7

Desktop Videoconferencing in Utah: The UPLIFT Project

Martin Agran
University of Northern Iowa
Richard Kiefer-O'Donnell
Minnesota State University

Title of Program

The Utah Consortium for Low Incidence Faculty and Teachers (UPLIFT)
University of Utah/Utah State University

Brief Description of Program

Overview of program. The Utah Consortium for Low Incidence Faculty and Teachers (UPLIFT) was a federally funded distance education personnel preparation project designed to ameliorate the critical teacher shortage in the area of severe disabilities in the largely rural intermountain and Northwest areas by providing pre- and in-service instruction.

It involved a multi-university consortium of universities (i.e., University of Alaska, Anchorage; University of Colorado, Colorado Springs; University of Northern Iowa; University of Utah; Washington State University) and two school districts in Utah (i.e., San Juan School District, Uintah School District) across five states (i.e., Alaska, Colorado, Iowa, Utah, Washington). The project utilized a desktop videoconferencing system designed for the use of the Internet that permitted two-way, real-time audio and video interaction between the instructor and all participating students. Additionally, the project had a web site that allowed students to retrieve readings, advanced organizers, and other instructional materials. The delivery of courses was via the transmission of courses to local receptor sites (e.g., university classroom, school district meeting room).

Students in the program received the same certification program they would have received on campus at the University of Utah or Utah State University (Note: Although certification requirements varied across consortium members, UPLIFT specialization courses exceeded the requirements of all consortium members). The program was offered at both the undergraduate and graduate levels. Based on their own instructional needs or the certification requirements in their states, students opted to complete the full program or take classes selectively. The certification program was comprised of five specialized didactic classes and two structured field experiences (Note: Required core or foundations classes were not included in the UPLIFT course sequence but were available via other distance education programs, school district in-service, or on campus from the host university). The didactic classes included Effective Service Delivery Models, Curriculum and Instruction, Positive Behavior Support, Collaboration and Management of Learning Environments, and Transdisciplinary Team and Instructional Approaches. The field experiences included one practicum experience and student teaching, both of which were supervised by a local field supervisor. The didactic classes were delivered via the Internet and required basic Internet use and computer skills of the students. At each receptor site, a technical facilitator was present.

The two co-principal investigators served as the primary instructors. In addition, a national faculty of experts served as adjunct faculty members. Faculty members included Dr. Martha Snell, University of Virginia; Dr. Fred Spooner, University of North Carolina – Charlotte; Dr. Fredda Brown, Queens College; Dr. Diane Baumgart, University of Idaho; Drs. Rob O'Neill and John McDonnell, University of Utah; Dr. Ed Helmstetter, Washington State University; Dr. Michael Giangreco, University of Vermont; and Dr. Stephanie Peck, Idaho State University. This national faculty allowed students to access state-of-the-art information from leading experts in the field. Faculty members delivered instruction from their home universities via the Enhanced CU-SeeMe software, and provided feedback and technical assistance to the project as needed.

Area(s) of Specialization

Low Incidence Disabilities

APPLICATION

Certification

PROGRAM LEVEL

Undergraduate and graduate pre- and in-service personnel preparation

TARGET AUDIENCE

Teachers in five states (i.e., Alaska, Colorado, Iowa, Utah, Washington)

ONLINE COMPONENT

Didactic classes delivered via the Internet, field experiences supervised by local field supervisors

TECHNOLOGY APPLICATION USED FOR ONLINE INSTRUCTION

COURSE MANAGEMENT SYSTEM/SOFTWARE AND HARDWARE

There were several interesting and unique aspects of the timing of UPLIFT project relative to the selection of technology. First, UPLIFT started without having a strong historical association with any one-course platform. Neither USU nor the University of Utah had a required platform, so the principal investigators were able to choose one that matched the instructional needs of the courses, faculty, and students. Then, given that the project operated with early or beta versions of course platform software, a website was needed as a back up for the synchronous tools of the course platform. Last, the project began during what was the "heyday" of new video conferencing software. There were over 100 different conferencing applications on the market. Sometimes they were very proprietary and required their own cameras and video capture cards. Rarely did they have good documentation and were often little more than beta products. They were seldom designed to meet the United Nation's International Telecommunication Union (ITU) H323 standards for being cross-platform or inter-operability. As such, project staff spent quite a bit of time early on learning and testing software. They had to change products because vendors went out of business, or the product was found to be less than robust.

Asynchronous Online Components

Presentation of content. The UPLIFT Course Management System employed a commercial course platform and a local website to deliver and manage all classes. Learning Space, sold first by Lotus, Inc, and now by IBM, Inc., was purchased and installed. It was chosen because of the scope of synchronous tools that it had available. Each was equally accessible by PC and Mac browsers which, in the mid-1990s, was a major need in that students used both platforms. While Learning Space had the capacity to hold documents for asynchronous access, early versions of the platform required use of Lotus Notes to manage the platform. It was felt that this process was too cumbersome for faculty. In its place, the website served as a resource to the student to retrieve course documents and hold, as needed, materials used during the class, if for whatever reason, the synchronous tools of Learning Space were not accessible.

Synchronous Online Format

Presentation of content. Graduate course delivery by UPLIFT was always synchronous. As many as six sites, each having 1 to 4 students from across the Intermountain West, met with one or two faculty instructors for nearly 3 hrs once a week. Given the location of the faculty, courses spanned 3 to 5 time zones. Commonly, one student per site per course received training in the operation software and login procedures. They handled troubleshooting locally. Each site had at least cable modem, DSL, or T1 access to the Internet. Students first would open the video conferencing software, connect to the conferencing server, and enter the course room designated for the evening session. After completion of sound checks and adjustments of cameras, the same student then used a browser to go to the course platform URL. There, a scheduled class forum was entered so as to access an array of synchronous tools used by the Instructor. Table 1 summarizes the various tools used by the project for instruction.

The synchronous tools of Learning Space were critical to the delivery of courses. Class periods were scheduled and documents posted to a whiteboard so that they could be seen and manipulated by all participants in the session. This served as the primary means to display information. Blank whiteboards were used for interaction, along with Application Sharing, a tool that permitted all users to see and modify any given application and file opened by the instructor. Data from a baseline activity, for example, were individually entered by students in class into the Instructor's spreadsheet, converted to linear graphs, and displayed to all students. It was then saved and emailed to all students within seconds of being created. Guided Web Browsing (called "Follow Me" by Learning Space) is a tool that let the instructor take all students to a new website. It could be one that had content associated with the session (e.g., information on augmentative communication devices, model projects for positive behavior support or para-educator training, etc.) or one with links to media on the streaming video server.

Both host universities had a strong history with the in-state ISDN or compressed video systems. A decision not to use these resources was made in that they were costly, sometimes difficult to access and schedule, and available only to Utah students and faculty. As a result, the project purchased and installed a 10-client Meeting Point Conferencing Server at the University of Utah. This server was ITU compliant for H 323 standards for inter-operability and cross-platform needs. It also permitted users to see up to nine other participants at once, bundling audio and video streams of all users to maximize and control for bandwidth consumption. Overall bandwidth was monitored regularly to ensure that the project did not overwhelm networks. Videoconference rooms were established for large group meetings, smaller breakouts, or 1:1 advising with students. The project used CUSeeMe (versions 1 through 4) to join the conference server room. Once done, students configured a "split-screen" layout of the VC window and the one from their Learning Space browser to display both the videoconferences images and whiteboards. New windows, created through Guided Web Browsing or from streamed video, were stacked temporarily above this layout.

FACTORS INFLUENCING DECISION TO USE ONLINE INSTRUCTION

Needs Addressed by Online Instruction

A critical shortage of qualified special educators—particularly those who serve students with severe disabilities—continues to be a major problem. Distance education provides a means for an individual interested in seeking certification or additional training who would otherwise be unable to access it to receive quality instruction Distance education technology represents a potentially effective means to ameliorate the teacher shortage problem. The benefits of such technology are obvious. Distance education literally puts education within research of persons interested in pursuing professional teacher certification but have been unable to do so it because of financial, logistical, and personal reasons. Distance education allows students to receive high quality accredited teacher education programs at convenient community sites or literally in their own homes. The associated costs related to the transmission of courses are relatively inexpensive.

Reasons for Selection of Specific Online Formats

University teacher preparation programs need to ensure that all educators leave with skills so they can best serve all students. For the latter, school districts, with or without university support, need to ensure that their teachers can best serve all learners. In both cases, distance education can meet this need. The Internet allows for extensive sources of information not easily accessed by other means. Distance education—in particular, internet-based distance education programs (e.g., UPLIFT)—permits the dissemination of instruction relating to inclusive practice that may not otherwise available. For both student teachers and teachers, classes or in-service workshops can be set up with experts throughout the country or faculty at a state university that is not within close proximity. Many of the problems associated with continuing education and professional development, such as travel, available time slots for satellite broadcasting, or one-way transmission systems, are eliminated.

Also, the live two-way audio and video capability of videoconferencing technology provides a potentially useful tool for practicum supervision required in certification programs. The observation of practicum students presents several challenges to teacher educators. Depending on the number of practicum students to be monitored, practicum supervision presents several geographical and logistical problems. To address this problem, videoconferencing allows for the live observation of practicum students from a distance. The advantages of this are several. First, feedback can be delivered immediately. Second, problems associated with the intrusiveness of a supervisor in a classroom are eliminated (over time, the reactive effects of a camera are reduced). Third, the technology serves as a valuable resource for problem solving and teaming among the student, the cooperating teacher, the faculty member, and others involved in the process (e.g., principal).

Last, distance education can serve to support teachers in their efforts to include all students. Distance education and other internet-based applications allow one to do this electronically. For example, if a teacher has a concern about a student's challenging behavior or how to set up an appropriate cooperative learning group, access to videoconferencing would allow this teacher to link up with specialists or other colleagues, both within and beyond the local area. For teachers in rural areas who are physically isolated from colleagues and other supports (e.g., itinerant specialists) and who have literally no one to talk to when a problem arises, the value of distance education is obvious.

OUTCOME AND IMPLICATIONS OF ONLINE INSTRUCTION

IMPACT OF ONLINE INSTRUCTION ON THIS PROGRAM

Over 70 students successfully completed the majority of the six UPLIFT courses. Slightly over 60% resided in Utah, with others coming from Colorado, Washington, Idaho, and Alaska. Of these states, only Colorado students resided in an urban setting that potentially gave them access to equivalent on-campus courses or resources. At the time, installation of needed software and audio-video equipment cost less than $250.00 per site. Once done, all access to courses was free for the site. Currently, these same components with faster USB connections cost less than $100.00. Total server costs were less than $5,000 though the events of 9/11 have since driven VC server costs up substantially.

Opportunities/Advantages/Benefits of Online Instruction

Students in the five participating states had immediate access to National Faculty members. Often these instructors taught using new resources from federal research or model projects that were under funding. The core model for delivery has since been replicated twice, in Montana and Virginia, to connect rural students to courses and faculty in severe disabilities. Both projects reported success.

Problems/Disadvantages/Limitations of Online Instruction

In many ways, UPLIFT functioned as a Research and Development project. The early 1990s saw a great influx of technology that had clear instructional potential. This provided the principal investigators an opportunity to "pick and match" those that appeared to meet the pedagogical needs of the courses with the instructional needs of the students. Oftentimes though, what was "reported on the box" didn't match what was found inside. Similarly, UPLIFT did not anticipate the time and resources needed to prepare students as active learners at-a-distance or faculty to teach fluidly with these technologies. Rarely did instructors have time to practice sufficiently to be fluent in concurrently using the different software. Likewise, preparation demanded greater time on their part, with most soon realizing that a "80-slide" PPT presentation for the evening session unrealistic and unsuccessful in meeting class goals.

SUGGESTIONS OR RECOMMENDATIONS FOR OTHERS CONSIDERING ONLINE INSTRUCTION

Within a year of initial implementation, it became clear that instructors had to be trained to promote interaction within the class. As one faculty member soon realized, "you can't see the raised eyebrows" of even the students sitting close to the camera. While UPLIFT staff informally promoted more interactive class session design formats, it was not fully successful in training faculty to implement these strategies reliably. Likewise, it became clear that more effort was needed to teach students to interact locally or across this virtual class environment. Additional training of faculty and students represents a major need for future implementation of the model.

LIST OF RESOURCES

BOOKS OR ARTICLES

Agran, M., & Kiefer-O'Donnell, R. (2001). Utah Consortium Project to prepare teachers of students with severe disabilities. In B. Ludlow & F. Spooner (Eds.), *Distance education* (p.173). Reston, VA: Council for Exceptional Children.

Smith, S., (2003), Online video conferencing: An application to teacher *education. Journal of Special Education Technology 18*(3),62-74.

Spooner, F., Agran, M., & Kiefer-O'Donnell, R. (2000). Distance education in the age of inclusion: Using telecommunications technologies to meet professional and personnel needs. *Journal of the Association for Persons with Severe Handicaps, 25,* 92-103.

Venn, M. L., Moore, R. L., & Gunter, P. L. (2001). Using audio/video conferencing to observe field-based practices of rural teachers. *Rural Educator, 22*(2), 24-27.

Web Sites

Teaching with video conferencing http://www.powertolearn.com/articles/teaching_with_technology/video_conferencing_for_teaching_and_learning.shtml

Contact Information

Martin Agran, Ph.D.
Professor and Department Head
Department of Special Education
University of Wyoming
McWhinnie 222
1000 E. University Ave.
Laramie, Wyoming 82071
307-766-2082 (phone)
307- 766-2018 (fax)
magran@uwyo.edu

Richard Kiefer-O'Donnell, Ph.D.
Associate Professor
Department of Education Studies: Special Populations
Minnesota State University, Mankato
318P Armstrong Hall
Mankato, MN 56001
(507) 389-5665
richard.kiefer-odonnell@mnsu.edu

Table 1

Technology Used by the UPLIFT Project

Tool	Product	Functions
Course Platform	IBM/Lotus Learning Space (versions 2 through 5)	1. Post and display MS PowerPoint class notes, Word documents, spreadsheets 2. Whiteboard for real-time individual or group writing during interactive assignments 3. Application sharing to mutually modify charts, forms or drafts within a spreadsheet or Word document 4. Guided web browsing, to take all participants to the other websites simultaneously 5. Later, the drop box used on limited basis for submission of assignments
Video conferencing	CUSeeMe (versions 1 through 4), along with their 10-client Meeting Point VC server (now marketed through First Virtual Communications)	1. Individual video conferencing by PC and Mac users into conference rooms that permit everyone to see each other at all times during the conferencing.
Course website	Created using Macromedia Dream Weaver; Adobe Acrobat (versions 2 through 4) to create PDF files	1. Access to PDFs for course readings, syllabi, assignments and handouts prior to class 2. Post PowerPoint and other documents as back-up to Learning Space

Tool	Product	Functions
Threaded Discussion	DISCUS server	1. Threaded discussions associated with class assignments
Streamed video	Processed with either Adobe Premier and Final Cut Pro	1. Posted initially on the web site as progressive download and then later on the University of Utah streaming video server so that video could be viewed and discussed while classes were also video conferencing.

Chapter 8

AN ONLINE SPECIAL EDUCATION ALTERNATIVE LICENSURE PROGRAM IN NEW MEXICO

TERESA ROWLISON
New Mexico State University

TITLE OF PROGRAM/SPONSOR

Special Education Alternative Licensure Program
College of Education at New Mexico State University

BRIEF DESCRIPTION OF PROGRAM

Overview of program. The special education alternative licensure program at New Mexico State University (NMSU) is a post-baccalaureate licensure program available totally online. The program, as approved by the New Mexico Public Education Department (NMPED), requires that candidates be individuals with a bachelor's degree who already are teaching with an internship license or substandard license (waiver). All of the courses, 21 credit hrs, were developed and are available online such that the program can be completed within one year. A Master's degree with an emphasis in special education is also available to program participants with the completion of an additional 15 credit hrs, which are available in a hybrid (online and face-to-face sessions) format.

Considering that program participants currently are teaching, it is felt that their teaching assignment provides extensive field experiences. The program requires that the employing district have an approved mentoring program by the NMPED. Integrated throughout the program are the issues of technology and multicultural education, which allows students to gain the skills and abilities to teach special learners with quality in New Mexico.

A brief description of the courses is as follows:

- **SPED 500** Survey of Programs for Exceptional Learners (ELs): This course provides characteristics, identification, and educational needs for ELs. Attention is given to the various types of programs serving ELs. This course is designed for all professional personnel who work with ELs.

- **SPED 504** Classroom Assessment of ELs: This course provides theory and use of norm- and criterion-referenced instruments in the classroom and planning of prescriptive instructional programs.

- **SPED 506** High Incidence Disabilities: This course introduces the high incidence disabilities in the field of special education.

- **SPED 507** Low Incidence Disabilities: This course introduces the low incidence disabilities in the field of special education.

- **SPED 509** Reading K-6 for ELs: This course provides reading strategies for ELs K-6.
- **SPED 511** Reading 7-12 for ELs: This course provides reading strategies for ELs 7-12.
- **SPED 559** Approaches to Classroom Management: This course provides behavior-change strategies for ELs.

Area of Specialization

Special education alternative licensure program K-12 offered as general program with area of specialization

Application

Graduate level certification and Master's degree program

Program Level

Inservice personnel preparation.

Target Audience

Persons with a bachelor's degree or persons who are already teaching on an internship license or substandard license (waiver)

Online Component

All of the courses, 21 credit hrs, have been developed and are available online.

TECHNOLOGY APPLICATIONS USED FOR ONLINE INSTRUCTION

Course Management System/Software and Hardware

This program utilizes the WebCT course management system for all courses on a server available to all faculty and students at NMSU as the main delivery mode. Course materials are created by each individual instructor using a variety of formats including HTML files, Word documents, or PDF files using Adobe Acrobat. Instructors may also use CENTRA and ITV as alternative or additional delivery modes. Dreamweaver, Flash, and Quicktime are available as supporting software. NMSU's Information and Communication Technologies (ICT) services provides professional development and technical assistance related to all of these formats and systems for both faculty and students using PC or Mac formats.

Asynchronous Online Formats

Presentation of content. Content in each course is designed by individual instructors and may include content modules or course assignments incorporating text, images, embedded audio or video clips, best practices, case studies, and links to appropriate Web sites. Many instructors also require that students purchase and read print resources, which may include textbooks, other professional reference books, or additional readings available online or through the NMSU library.

Interactions with learners. A community of learners is created through interactions with each learner and among learners as a group. This is accomplished primarily through WebCT tools such as mail, calendar, discussions, student homepages, course assignments, chats, and quizzes. Each instructor determines what students are required to participate in, which may include weekly discussion questions, weekly course assignments, group or individual projects, whole group or small group chats, etc.

All instructors are required to adhere to the NMSU College of Education (COE) Conceptual Framework, which ensures collaboratively developed education and clinical professionals who facilitate learning and are responsive to diverse and changing environments. The aim of the COE is to ensure that its students have opportunities to become creative and critical thinkers who can make appropriate decisions relative to their professional roles and responsibilities. Graduates of the NMSU COE will possess broad general education and content area knowledge, be effective and reflective practitioners and problem-solvers, apply innovative learning technologies, and participate in opportunities for professional growth. Through its efforts, the COE seeks to empower its graduates, enabling them to acquire the knowledge, skills, and dispositions that will allow them to excel in their institutional, clinical, or administrative responsibilities. In order to address, actualize, and assess the outcomes of program activity and student progress, the COE collaborates with educational professionals on campus, in the schools, and in the community. The goals of the courses in this program are consistent with the COE's Conceptual Framework. The program is designed to prepare effective practitioners who are reflective, who are sensitive to diversity, and who are able to work effectively with students with diverse needs.

Interactions with learners are also assessed through online course evaluations. These online evaluations ensure course effectiveness and improved learning opportunities for special education teachers and the children and families of New Mexico they serve. A database has been set up to allow students to anonymously submit course evaluations online. These evaluations are received by the Assistant Director for Distance Education for the COE, who summarizes them and provides feedback to the participating instructors.

Results of these evaluations for the 2003-2004 academic year indicate that the online courses within the special education program were moderately difficult (55%), required a reasonable amount of effort (95%), stated clear objectives (95%), required a textbook that was very helpful (86%), and had course materials that students were able to access (79%). The overall rating of instructors was excellent or very good (92%). Students described the ease of navigating and using the online materials as above average or excellent (63%). Students compared the value of these courses against other courses at the same level they had taken at a distance as above average or excellent (61%). Students compared the value of these courses against other courses at the same level they had taken in a traditional classroom setting as above average or excellent (75%). Students described the level of interaction with faculty and other students as average, above average, or excellent (76%). The most frequent comments students made were related to the online format of the classes and that they liked being able to work on their own schedule. Tabulated results of these evaluations are provided in Appendix A. Similar course evaluation data are being collected for the 2004-2005 academic year.

FACTORS INFLUENCING DECISION TO USE ONLINE INSTRUCTION

Needs Addressed by Online Instruction

In response to the critical shortage of special education teachers in New Mexico, the COE at NMSU developed the special education alternative licensure program online in order to increase the number of qualified special educators in the state. A legislative allocation during 2003 through Senate Bill (SB) 655 provided funds for one year to develop the program, which consists of seven online courses, to address the critical shortage of appropriately licensed teachers, service providers, and administrators. Enrollment in these licensure courses for the 2003-2004 academic year was 270. The seven licensure courses may be applied towards a Master's degree with an emphasis in special education. The remaining 15 hrs required for a Master's with an emphasis in Special Education are available in hybrid format. Enrollment in these licensure and Master's courses for the 2004-2005 academic year was 281.

Reasons for Selection of Specific Online Formats

Around 1996/97, NMSU was considered a beta tester of WebCT. NMSU was one of the first 69 universities who were testing WebCT in the world. Through demonstrations and word of mouth, faculty began to use WebCT. Within the

College of Education, there are 39 courses offered via WebCT this semester alone. Additionally, instructors may use alternative delivery modes and supporting software in their distance education courses. ICT provides training and technical assistance for faculty and students regarding the delivery modes and software programs available at NMSU.

OUTCOMES AND IMPLICATIONS OF ONLINE INSTRUCTION

Impact of Online Instruction on This Program

It is difficult to determine the number of special education alternative licensure program completers because licensure is issued by the NMPED, not NMSU. Also, there is no way of knowing how many students took one or two of the online classes to complete a licensure program from another institution. However, we do know that the statewide number of special education teachers on internship or substandard (waiver) licenses during the 2002-2003 academic year was 559 and has been reduced to 357 during the 2003-2004 academic year. Data are not yet available for the 2004-2005 academic year. This is a significant reduction in internship or substandard licenses (waivers) granted in the area of special education. Although, the online special education alternative licensure program is not solely responsible for this reduction, this program certainly contributed and continues to meet a need within the state.

Opportunities/Advantages/Benefits of Online Instruction

The opportunities of online instruction allow special

education teachers on internship or substandard licenses (waivers) from anywhere in the state to have access to the required course work for them to apply for their Level I teaching license. The advantages and benefits of online instruction include the ability to provide more individualized learning activities and flexible schedules for both students and instructors. Additionally, there is increased student responsibility for personal learning and instructor accountability for achievement of student learning.

Problems/Disadvantages/limitations of Online Instruction

One of the initial problems with online instruction was the lack of already trained faculty to effectively teach online. This resulted in a need for ongoing technical assistance for faculty while they were teaching online for the first time. Technical assistance was provided in the form of monthly program meetings where training on different WebCT tools was provided. These monthly program meetings were scheduled by the program coordinator. Topics were determined by the participants and included the assignment drop box, chat, discussion, and quizzes. The training was provided by ICT personnel, and a WebCT course was provided for distance education instructors to receive additional training and post discussion questions related to online teaching. Attendance at the monthly program meetings ranged from 5 online instructors to all 13 of the online instructors who taught during the first year of this program.

One of the disadvantages to online instruction is the additional effort needed on the part of the instructors and students to create a community of learners without face-to-face contact. Some of the limitations of online instruction include technical difficulties that may occur and the increased amount of time and effort needed to develop and deliver online instruction in comparison to the traditional face-to-face format.

SUGGESTIONS OR RECOMMENDATIONS FOR OTHERS CONSIDERING ONLINE INSTRUCTION

For others considering the provision of online instruction, it is essential that there be departmental and college support for such a course delivery format. It may be necessary to develop and demonstrate the course work to departmental and college faculty prior to offering the course to increase awareness and acceptance for online classes. It is also important to determine the financial support for online course development. The COE at NMSU provides an additional one-time stipend as compensation for instructors to develop online courses. Instructors are also a portion of their 9-month salary in additional compensation when they teach online courses as an overload. It also is recommended that there be a cap on enrollment for online courses, such that high quality instruction is provided. Most of the online courses at NMSU are capped at an enrollment of 20.

It is suggested that a program coordinator be hired. Program Coordinator responsibilities should include facilitating the development of the courses, providing technical assistance to instructors, developing program brochures and newsletter for dissemination, traveling around the state to promote the program, attending state-level meetings and conferences, presenting at state and national-level conferences, and other activities to ensure project sustainability.

Technical assistance must be readily available to faculty and students. The distance education office in the COE at NMSU has developed handouts for getting started for students. ICT provides training for faculty, which is paid for by the distance education office in the COE for instructors teaching special education alternative licensure and Master's courses. Additionally, a student tutorial has been developed by the NMSU Office of Distance Education. Many instructors also provide a link to the WebCT student tutorial within their online classes. Technical assistance also is provided to student and faculty on an individual basis by the COE Assistant Director for Distance Education and the special education alternative licensure program coordinator.

Finally, flexibility is paramount. Individualized accommodations must be provided to students who experience technical difficulties. Fortunately, in New Mexico, all of the public schools have Internet access and many have broadband. However, not all of the students participating in this online course work are able to access their courses from the school in which they work due to time constraints and/or school policies.

LIST OF RESOURCES

Books and Articles

Losco, J., & Fife, B. L. (Eds.). (2000). *Higher education in transition: The challenges of the new millennium.* Westport, CT: Bergin &

Garvey.

Russell, T. L. (1999). *The no significant difference phenomenon as reported in 355 research reports,*

SUMMARIES AND PAPERS

A comparative research annotated bibliography on technology for distance education. North Carolina State University: Office of Instructional Telecommunications.

Van Dusen, G. C. (2000). *Digital dilemma.* San Francisco, CA: Jossey-Bass.

WEB SITES

http://education.nmsu.edu/prospective/Licensure/licensure_endorsements.php
http://ict.nmsu.edu/
http://distance.nmsu.edu/sop/index.html
http://www.webct.com
http://www.ion.illinois.edu/IONresources/tutorials/webct/mvcr/flash.html

CONTACT INFORMATION

Teresa Rowlinson, Ph.D.
Assistant Professor/Program Coordinator
New Mexico State University
MSC 3R/P. O. Box 30001
Las Cruces, NM 88003-8001
(505) 646-2125
Rowlison@nmsu.edu

Appendix A
Evaluation Instrument

NEW MEXICO STATE UNIVERSITY
COLLEGE OF EDUCATION-DE PROGRAMS
Course/Instructor Evaluations
Fall 2003 and Spring 2004

Course	Section	Instructor	Location	Respondents
SPED	All	All	Online	84

Course

Question	Responses	Distribution		
Subject	84	Not difficult at all	Moderately difficult	Very difficult
		38	46	0
Effort	84	Too low	Reasonable	Excessive
		0	80	4
Objectives	84	Not clear	Somewhat clear	Clear
		0	4	80
Textbook	84	Not helpful	Somewhat helpful	Very helpful
		3	9	72
Materials	84	No materials required	Access to materials was difficult	I was able to access the materials
		8	10	66

Instructor

Question	Responses	Distribution				
		Poor	Fair	Good	Very Good	Excellent
Knowledge	70	0	1	4	22	43
Convey	67	1	0	13	16	37
Stimulate	70	1	1	13	24	31
Organization	76	0	1	8	20	47
Grading	75	0	1	4	21	49
Attitude	76	0	0	5	13	58
S-I Interact	75	2	1	14	14	44
Accessibility	76	1	2	12	20	41
Overall	84	1	0	6	18	59

WRITTEN COMMENTS

Examples:

Liked being able to do work on my schedule.

Easy access, not having to leave home.

Liked being able to sign in and do classwork when it is convenient for me.

I love the fact that I can work on it when I have time, without having to attend a class.

Liked being able to work at your own time.

It is more convenient than the traditional classroom.

Chapter 9

LIVE WEBCASTING: AN INTERACTIVE ONLINE PROGRAM TO PREPARE PERSONNEL FOR RURAL AREAS

BARBARA L. LUDLOW
West Virginia University
MICHAEL C. DUFF
Discover Video Productions

TITLE OF PROGRAM/SPONSOR

SMD/EISE Distance Education Program
College of Human Resources and Education at West Virginia University

BRIEF DESCRIPTION OF PROGRAM

Overview of Program. The Severe/Multiple Disabilities and Early Intervention Special Education (SMD/EISE) Program at West Virginia University (WVU) is a post-baccalaureate certification and degree program that has been delivered entirely at a distance since 1990. Initially delivered via satellite broadcasts, the program later converted to half satellite-half Web format, and is currently available entirely online to students anywhere in the world. For courses, the program uses a combination of live webcasts transmitted in real time with telephone interaction between students and instructors and other materials and activities available online to be completed at the student's convenience. For practicum experiences, students may use their own job setting if employed as a special educator in an appropriate placement or they may arrange for a placement in a local public school or community agency. The program features 11 courses and 4 practicum experiences that are offered across a 3-yr program cycle, with four courses offered per year. Students are required to complete eight courses and six credit hrs of practicum for certification plus two elective courses for a master's degree in one area plus an additional three courses and three credit hrs of practicum for a second certification. The program is targeted at practicing but unlicensed special educators working in West Virginia and the neighboring states, but it is also available on a limited basis to students in other states and in international locations.

Area(s) of Specialization

Severe/Multiple Disabilities
Early Intervention/Early Childhood Special Education

Application

Post-baccalaureate certification and/or Master's degree program

Program Level

Preservice personnel preparation

Target Audience

Practicing but unlicensed educators working in rural schools and agencies throughout the United States (and in selected international locations)

Online Component

All courses are offered entirely online; practica are offered in local areas with local supervision with online support from university personnel; learners are not required to come to campus at any time

TECHNOLOGY APPLICATIONS USED FOR ONLINE INSTRUCTION

Course Management System/Software and Hardware

This program utilizes the WebCT course management system for all courses and practicum experiences on a server available to all faculty and students at WVU. Course materials are created as HTML files in Macromedia Dreamweaver, interactive exercises are designed in Macromedia Flash, and forms that need to be printed out are formatted as PDF files using Adobe Acrobat. Webcasts are delivered using Apple QuickTime 6.0 streaming video; Apple QuickTime Broadcaster is used to create two separate streams (one at 56 kbps for dial-up telephone modems and another at 128 kbps for cable modems and digital subscriber or T1 lines); and streams are hosted on an Apple Macintosh OS10 streaming server. Students who live in remote rural areas where 56 kbps access is not currently available are permitted to participate in live webcasts through Wimba for WebCT, an online audio conferencing program; these student are able to hear (but not see) the presentations and interactions and interact with the instructor and other students through that format as well.

Asynchronous Online Formats

Presentation of content. Content in each course is presented through eight content modules incorporating text, images, embedded audio clips of interviews with professionals and family members, video clips of best practices, case studies, and links to Web sites. Each module contains three sections that open with an experiential activity that asks them to reflect on their prior knowledge and closes with an application activity that asks them to apply the information they have just learned to a specific case. Students also are required to purchase and read print resources, which may be textbooks or other professional reference books, and to access additional readings as needed through the WVU online library resources.

Interactions with learners. Interactions with each learner and among learners as a group are accomplished primarily through WebCT formats, such as mail, discussion boards, and assignment and quiz tools. All students are required to participate in discussions related to eight discussion questions posed by the instructor. Discussions require recounting personal experiences related to the topic, sharing effective strategies, searching for resources online, locating articles in the online library, brainstorming solutions for a problem in implementing best practice, presenting solutions to a case study, discussing concerns about professional duties, and debating controversial issues. Students are assigned to discussions in groups of 6 to 8 students based on the age/grade level of interest to them. Some courses and the practicum experience also make use of the Horizon/Wimba for WebCT voice board, where students record and post voice messages for discussions. In addition, each student completes a multiple-part applied project applying course principles and practices in a real world setting that serves as a field experience associated with the course. Some courses include other activities such as position papers or collaborative group activities. Finally, students prepare two case-based essay exams as a mid-term and final assessment of their understanding of course content.

Synchronous Online Formats

Presentation of content. Content in each course is presented by means of short lectures, demonstrations, video clips, and live or taped guest interviews accompanied by screen graphics in five live webcast sessions, each of 2 hrs duration.

Interactions with learners. Interactions with each learner and among learners as a group are accomplished by means of a telephone conferencing system operated by the state which students access using a toll free number and passcode for the course conference. The instructor can call on specific students or ask several students to interact on the phone in discussions, debates, role-plays, simulations or case studies. Students are provided with information needed to complete in-class activities in advance as Portable Document Format (PDF) files in posted in the course which they print out in advance of each class session and fill out before or during activities. At the end of each class session, students are required to complete a documentation activity as evidence of their attendance; this activity involves submitting a mail message within 20 min with a response related to a question posed by the instructor related to specific content topics. Some live interactions also are conducted using the WebCT Text Chat tool or the Wimba for WebCT Voice Chat function.

FACTORS INFLUENCING DECISION TO USE ONLINE INSTRUCTION

Needs Addressed by Online Instruction

West Virginia and the states in the surrounding Appalachian region have experienced and continue to experience chronic shortages and high attrition rates of qualified special education personnel. This situation leaves rural school systems faced with the challenge of continuing recruitment of local people into unstaffed positions as well as the need to help them access personnel preparation programs to become fully qualified and licensed to teach. The SMD/EISE Program at WVU has offered a distance education program to practicing but uncertified special educators in the state and region since 1990 using the state's satellite network. In 2000, the program responded to requests from individuals living in areas outside the region with no access to satellite downlink sites as well as to individuals living in international areas outside the satellite footprint by adding a webcasting component to the program that was accessed by those students only. In 2003, the university was informed that the state would no longer support the satellite network and that distance education programs needed to seek option delivery options. The webcasting format allows the SMD/EISE program to continue to serve its current target audience with no disruption of service or change in program components. It also will allow the program to further expand its outreach efforts to rural areas across the country as well as to individuals living and working in other parts of the world (e.g., U. S. protectorates, Department of Defense Schools, or American embassies).

REASONS FOR SELECTION OF SPECIFIC ONLINE FORMATS

The WebCT course management system was initially selected by WVU because it is a comprehensive package of teaching and learning tools with many desirable features; the state's Higher Education Policy Commission now requires all colleges and universities to use WebCT under a statewide licensing agreement. The SMD/EISE Program has chosen to use QuickTime streaming products because this streaming format is compatible across platforms, produces the highest quality video and audio signal at all data rates, and features free software and low cost hardware that is inexpensive and simple to set up and operate.

OUTCOMES AND IMPLICATIONS OF ONLINE INSTRUCTION

Impact of Online Instruction on This Program

The SMD/EISE Program has used all online instruction for some students (approximately 10-12) since 2001 with considerable excellent outcomes and made a successful transition to all online instruction for all students (over 100) with two courses and both practicum experiences in Fall 2004. All 11 courses and both practica in the program will be converted to all online instruction by Spring 2007. This transformation has not only increased interest in the program on the part of individuals residing outside the state (which is likely to result in an increase in enrollment), but it also positions the program to alter its existing program cycle (offering each course once every three (3) years) to allow more frequent course offerings and speedier completion of certification requirements.

Opportunities/Advantages/Benefits of Online Instruction

The major benefits of online instruction include enhanced program access to prospective and practicing special educators, especially important in rural areas; higher enrollments with lower delivery costs for low incidence programs; improved quality of instruction through greater use of multiple media and more individualized learning activities; and increased student responsibility for their own learning and instructor accountability for achievement of learning outcomes.

Problems/Disadvantages/Limitations of Online Instruction

The major drawbacks of online instruction include greater effort needed on the part of instructors and students to establish meaningful relationship with the loss of face-to-face contact, more instructor time and effort needed to develop and deliver instruction, and stress caused by technical problems.

Suggestions or Recommendations for Others Considering Online Instruction

In making the decision to move to online instruction in part or in whole or to use synchronous or asynchronous formats or both, programs need to consider questions related to access, technology and technical support availability, and cost. First to address is the question of who is the target audience for the program and what Internet access do they have; some rural areas have only low bandwidth access or connection problems stemming from severe weather conditions or frequent power outages that would make applications such as streaming video unrealistic. Unless students have reasonably reliable access to online instruction and learning activities, they will drop out of courses, frustrate faculty with complaints, or express dissatisfaction on evaluations. A second concern is what hardware and software is available for use by instructors and students; colleges and universities generally use a specific course management system or collection of Web applications, which may constrain the type of activities that can be utilized, while students may have computers with limited capacity, incompatible browsers, or a restricted range of software that will prevent them from accessing some online formats. Whenever possible, programs should make use of hardware that is already at hand and simple to use and software that is widely available or can be purchased at low cost or available as a free download. Next, programs should explore what training and support will be available for specific formats, both for instructors and for students, during the development and initial learning phases as well as on a continuing basis. All participants in online instruction need help in learning to teach and/or learn successfully and assistance in solving technical problems as they arise. Finally, some consideration should be given to the costs of the proposed system from the perspective of the program and the student. Simple online communications can be relatively inexpensive, while complex course management systems can be quite expensive for the institution; some video streaming applications can be set up using basic programs on any computer for almost nothing, while others require purchase of proprietary systems costing thousands of dollars per distant site.

LIST OF RESOURCES

Books and Articles

Austerberg, D., & Starks, G. (2002). *The technology of video and audio streaming.* Burlington, MA: Focal Press.

Ludlow, B. L., & Duff, M. C. (2002). Lice broadcasting online: Interactive training for rural special educators. *Rural Special Education Quarterly, 21*(4), 26-30.

Mack, S. (2002). *Streaming media bible.* New York: Hungry Minds.

Sauer, J. (2001). The stream team: A digital video format primer.

Emedia, 14(10), 36-41.

Topic, M. (2002). *Streaming media demystified*. New York: McGraw-Hill.

Waggoner, B. (2001). Web video codecs compared. *Digital Video, 9*(11), 22-36.

WEB SITES

http://www.webct.com
http://www.apple.com/quicktime

CONTACT INFORMATION

Barbara L. Ludlow, Ed.D.
Professor and Program Coordinator
SMD/EISE Program
608 Allen Hall
West Virginia University
Morgantown, WV 26506-6122
304-293-3450
bludlow@mail.wvu.edu
http://www.smdeise.wvu.edu

Chapter 10

Desktop Videoconferencing for Practicum Supervision: Two-Way Interactive Video and Audio with Polycom Viewstation 128

Jane B. Pemberton
Texas Woman's Universityy
Tandra Tyler-Wood
University of North Texas
L. Nan Restine
Texas Woman's University

Title of Program/Sponsor

Special Education Program
College of Education at the University of North Texas

Brief Description of Program

Overview of Program. The Educational Diagnostician Program in the Department of Technology and Cognition at the University of North Texas (UNT) is a certification and Masters' degree program that has been delivered to students on the main campus in Denton, TX, and at a distance to the UNT System Center in south Dallas since 1998. Additional sites in rural areas are established based on students' location needs. All didactic classes for the Educational Diagnostician Program are delivered either through a Tandburg system or web-based through the Internet using the WebCT course management system. Tandburg is a two-way interactive closed-circuit television system that uses fiber optics, microwave, phone lines, and a digital satellite to deliver educational information to selected sites. WebCT is used at UNT to support students and their professors during the delivery of didactic courses. Some courses are delivered totally over WebCT, while others are with WebCT support, along with the Tandburg system. The Educational Diagnostician Program is designed for students seeking a Masters' degree and/or certification as an Educational Diagnostician. The program requirement for a Masters' degree is 36 to 42 credit hrs, depending upon the student's previous coursework and experience. For students that have a Masters' degree and are seeking certification only as an Educational Diagnostician, the course requirements are 12 to 30 credit hrs, again depending upon the student's previous coursework and experience. Each course is offered one to two times a year. A state exam at the completion of the program is considered one of the most difficult certification exams in the state of Texas. Therefore, students must pass a series of benchmarks built into key courses in order to meet the criteria to sit for the exam. At this time, UNT has a 100% pass rate at the state level for certification. There are three courses in the Educational Diagnostician sequence (EDSP 5100, EDSP 5530, and EDSP 5540) that require field-based or practicum work. Prior to practicum experience in these three assessment classes, students must "check out" with one-to-one testing between the university student and the professor. For example, in EDSP 5100, a student must demonstrate skills in administering a standardized assessment, such as the *KeyMath-Revised* or *Woodcock Reading Mastery Test-Revised*. For university students at a distance, the "check out" component of the program can be accomplished through the Polycom system. For actual practicum experience in the assessment classes, university students arrange for individuals to be assessed that are not currently in special education nor considered for special education. This supervision also can be accomplished through desktop videoconferencing. The remote supervision option for desktop videoconferencing was field-tested by students and faculty between two faculty offices, in order to establish protocols and methods for actual supervision in the future using standardized assessment materials. In addition to standardized assessments, students complete curriculum-based assessment (CBA) assignments. The Polycom system is used to supervise students with the CBA assessment, while WebCT is used as a vehicle for turning in and receiving feedback on assignments. The use of Polycom and WebCT allows students to receive timely feedback. Collaboration between the faculty member and university student, as well as direct observation of the university student collecting assessment data, can be accomplished using the Polycom ViewStation 128 connection.

AREAS OF SPECIALIZATION

Educational Diagnostician Program

APPLICATION

Masters' degree program and/or post-Masters' certification

PROGRAM LEVEL

Graduate-level personnel preparation

TARGET AUDIENCE

Educators working in rural and urban schools that are seeking certification as an Educational Diagnostician. The Educational Diagnostician is a key team member in schools who provides eligibility information for Special Education that must be considered in order for a student to receive services under the Individuals with Disabilities Act (IDEA). The audience for remote supervision, particularly with CBA projects, was extended in 2005 to students enrolled in the Special Education Masters' Program at Texas Woman's University (TWU). Both UNT's and TWU's main campuses are in Denton, TX.

ONLINE COMPONENT

All courses are offered either online through WebCT or through the closed circuit Tandburg system. Evaluating the use of assessment tools prior to practicum experiences for students at a distance is accomplished either one-on-one with the faculty member teaching the course or through a Polycom ViewStation 128. Learners are not required to come to the main campus in Denton at any time. Arrangements are made for broadcasting through the Tandburg system to rural sites prior to semester classes beginning.

TECHNOLOGY APPLICATIONS USED FOR REMOTE SUPERVISION

Course Management System/Software and Hardware

This program uses the Polycom ViewStation 128 to deliver field-based supervision opportunities to students in areas outside the main campus of the university. Information on the Polycom unit can be found at www.tribecaexpress.com/polycom_viavideo.htm. The Polycom ViewStation 128 includes the following: (a) document camera input, (b) VCR recording and playback, (c) telephone input, (d) 10 megabit LAN connections, (d) optional 2nd microphone, (e) ISDN or Ethernet/LAN connections, (f) data speeds between 128 kbps and 768 kbps, and (g) optional upgrades for quad BRI and optional upgrades for multipoint communications. In addition, the ViewStation 128 provides full motion video (30 fps), echo cancellation, and allows the remote site to control both cameras, including tilt, zoom, and plan functions. The desktop videoconferencing system is real time (synchronous) and enables a university student and faculty member to see and hear each other. Participants can have conversations at a distance. The Polycom ViewStation uses the high bandwidth available at the university, which is a T1 or higher connection, with no added overhead. End-users (schools) with DSL and/or cable modem connectivity can cruise at comparable, but not guaranteed, T1 speeds. Although T1 and better connections (T3, OCR3, etc.) offer 1.5 Megabits per second (Mbps) and higher constant transmissions, DSL and cable modem offer the end-user higher Mbps for upstream transmission and about one-third of that speed for downstream transmission. Three major factors affect the rate for transmission for DSL and cable modem use: hardware capabilities of the modem, bandwidth allocated by the Internet Service Provider (ISP) or the user, and traffic on the network. The ViewStation connects directly to a television and does not require a computer. Based on our experience with a different desktop videoconferencing system (EnVision with Sorenson), some individuals in the schools were uncomfortable installing computer hardware. The ability to connect to a standard television set seems to greatly reduce the technical expertise needed at the remote location. For this project, the Polycom is placed on top of a television set and a two-inch wide, industrial-strength Velcro is attached to the top of the television and the bottom of the Polycom unit to keep the unit from falling.

Synchronous Online Formats

Interaction with learners. Polycom ViewStations and television sets are checked out to university graduate students enrolled in the Special Education Program. Typically, permission to set up the Polycom goes through the school principal and the technology support person for the district. There are two Polycom units in key faculty members' offices on the main campus and one unit available in the general access lab located in the education building. Schedules are set for supervision, advising, collaboration, and other uses. Students and faculty are encouraged to use the systems in spontaneous, practical ways. For example, two special education teachers in one rural area used the Polycoms for conversations about class assignments, social events or just chatting. At the beginning of the semester, the teachers were hesitant to relate to the faculty member that the primary use of the Polycom was the social connection between the two teachers. In effect, the beginning informal uses helped the teachers feel more comfortable using the Polycom with the professor of the class when needed for supervision, advising, or collaboration.

FACTORS INFLUENCING THE DECISION TO USE DESKTOP VIDEOCONFERENCING

Needs Addressed by Desktop Videoconferencing

There is a critical shortage of certified Educational Diagnostician in Texas and other areas of the United States. This shortage is often addressed by school personnel taking positions as Educational Diagnosticians in rural areas prior to receiving certification. Consequently, individuals seeking an Educational Diagnostician certificate must take classes while working full-time, making time and travel requirements critical factors in their decisions to pursue additional coursework. For the faculty member, supervision of students in a practicum setting during university training can result in additional time spent in travel, which can have a negative impact on the amount of time available for actual supervision of the university students. Attempts of videotaping the practicum experience or hiring an adjunct for on-site supervision are often regarded as less than satisfactory for the student and the faculty member.

REASONS FOR SELECTION OF SPECIFIC DESKTOP VIDEOCONFERENCING SYSTEM

The reasons for selection of Polycom ViewStation 128 came as a result of prior efforts to supervise students in rural areas through technology. The authors had previous experience with the Sorenson EnVision desktop videoconferencing system and found that system to work well once the connection was established. However, additional technologies became available for school use. As new technologies are developed, it is important to continue to field-test to help determine appropriateness for the needs of the students, faculty, and academic program. It is important that the selected system is affordable and practical for use in the schools. In addition, user support is a critical factor for many districts when seeking to implement new technology. The Polycom ViewStation 128 met the criteria for use.

OUTCOMES AND IMPLICATIONS FOR DESKTOP VIDEOCONFERENCING FOR SUPERVISION

IMPACT OF ONLINE SUPERVISION IN THIS PROGRAM

Faculty members in the Educational Diagnostician Program have used the Polycom ViewStation 128 extensively for collaboration between the two faculty offices. In addition, an earlier study conducted by the authors focused on rating student proficiency while observing through EnVision 28 university students. Each student administered selected sections from either the *Woodcock Reading Mastery Test-Revised* or the *Woodcock Johnson III Test of Achievement*. Both tests are standardized. The study took place between the two faculty offices, rather than in teachers' classrooms, in order to establish the feasibility of supervision at a distance. During the actual observations of each of the 28 participating students, agreement on a proficiency rating scale between the two observers was 85%. During a different study, Polycom ViewStation 128s were used with a CBA assessment in a reading fluency project. Students from Rose Street Day Treatment School, which is a partnership school between a medical facility and the Wichita Falls (Texas) Independent School District, were observed on-site by their teacher and at a distance by university faculty using the Polycom ViewStation 128. The teacher was working on her Masters' degree and certification as an Educational Diagnostician. Results from the Rose Street School study indicate satisfaction from the teacher, students, and faculty when measuring the accuracy of word count during the fluency assessment.

Opportunities/Advantages/Benefits for Desktop Videoconferencing for Supervision

Major benefits include the accessibility or ease of use with the desktop videoconferencing system that makes it possible to directly observe, supervise, and collaborate with students at a distance. Enlisting the expertise of other faculty members is also a plus, which permits the establishment of inter-related reliability when two faculty members share responsibility for observing and rating students in field-based settings. Immediacy, frequency, and accuracy are all factors that enhance communication, and in turn support the supervision process.

Problems/Disadvantages/Limitations of Desktop Videoconferencing for Supervision

Major drawbacks include actual set-up and use in the schools. The university setting has strong technology support, so any difficulties in set-up or use could be addressed quickly. In the school setting, the university student is frequently hesitant to address technology issues and often waits for the technology coordinator in the school to fix a problem, causing delays in use. In one school, squirrels chewed the connection so that attempts to collaborate were delayed until new wires could be installed. Based on our experiences, the technology appears to be still in the development stages for application on a wide scale use between universities and school districts for supervision.

Suggestions or Recommendations for Others Considering Desktop Videoconferencing for Supervision

Desktop videoconferencing is becoming more available and affordable. Online networks such as thefacebook are now found at 200 colleges and universities. Cell phones and PDAs have video components. As the technology continues to develop, supervision at a distance can become an integral part of the distance education program. If Polycom ViewStations are used along with a television, it is important to select a television that is easy for one person to transport from the university to the school. Some TVs are so heavy that mobility is compromised. A flat screen takes less room if placed on an office desk, and depending upon size is often more portable. Because the initial hook up was often an issue with new users, a step-by-step guide with color pictures and directions was distributed with each Polycom. In addition, contact support people at the university were available to talk a person through the set up. It was important to support the school with the initial set up, because once the link was made the technology was exciting for the user to use in meeting the requirements for the university program. When the user is motivated, frequency of use and maintaining connections seems to be much smoother.

LIST OF RESOURCES

BOOKS AND ARTICLES

Binner, K. (1998). *The perceived benefits and limitations of using two-way conferencing technology to supervise pre-service teachers in a remote teacher training program.* Unpublished masters' thesis, Utah State University, Logan, UT.

Ludlow, B.L., & Duff, M.C. (2001) Guidelines for selecting telecommunications technologies for distance education. In B. Ludlow & F. Spooner (Eds.). *Distance education applications in teacher education in special education.* Arlington, VA: Teacher Education Division Council for Exceptional Children.

Pemberton, J.B., Cereijo, M., Tyler-Wood, T., Rademacher, J. (2004). Desktop Videoconferencing: Examples of Applications to Support Teacher Training in Rural Areas. *Rural Special Education Quarterly, 25*(2), 3-8.

Wong, M. (2004, December 7) Face-to-face encounters take back seat to online social life. *Ft. Worth Star Telegram,* p. 11A.

CONTACT INFORMATION

Jane Pemberton
Texas Woman's University
Department of Professional Education
P. O. Box 425769
Denton, TX 76204-5769
940-898-2273
jpemberton@mail.twu.edu

Chapter 11

PROJECT IMPACT*NET: A DISTANCE IN-SERVICE MODEL TO INCREASE CLASSROOM SKILLS OF PARAEDUCATORS AND THEIR SUPERVISING TEACHERS

DAVID E. FORBUSH
ROBERT L. MORGAN
Utah State University

TITLE OF PROGRAM/SPONSOR

Project IMPACT*NET
Utah State University
Sponsored by Office of Special Education and Rehabilitation Services – H325N000048

BRIEF PROGRAM DESCRIPTION

Overview of Program. In January 2001, Project IMPACT*NET was launched with a grant award from the Office of Special Education Programs (OSERS). The objectives of the project were to (a) deliver training to paraeducators and to teachers nationwide using an Internet-based distance delivery system, (b) develop and implement a model for distance delivery of inservice training to enhance the skills of paraeducators working in special education and Title I settings, and (c) implement distance delivery of inservice training to improve the effectiveness of classroom teams composed of paraeducators and teachers. The overarching goal was development of a model for standardized, field-tested, in-service training program to enhance the instructional skills of paraeducators, and the instructional effectiveness of classroom teams.

Over a four-semester period, two Utah State University instructors using iVisit video conferencing software and the Internet delivered two separate courses to 133 paraeducators and 68 classroom teams of paraeducators and teachers. Participants included paraprofessionals and teachers in Colonial School District in Delaware, Madison School District in Idaho, Warrior Run School District in Pennsylvania, and Box Elder School District in Utah. Use of the instructional delivery system and iVisit software allowed participants at each site to receive training simultaneously and to hear and see the instructor and participants at other sites. Through this system participants were able to form collegial relationships and engage other participants at multiple sites.

The first course was developed for paraeducators working in Title I and special education programs. The second course was developed for classroom teams of teachers and paraeducators. Two existing curricula formed the primary instructional objectives and content for each course. Enhancing the Skills of Paraeducators, 2nd edition (ESP: 2) was used for the paraeducator class and Colleagues in the Classroom (CINC) was used for the classroom team course.

Site coordinators were hired for each distant class location to act as liaisons for the project and to provide supplementary instructional support to students (e.g., facilitate site-based discussion, grade assignments, and address assignment-related questions). Site Coordinators also assisted in setting up, launching, and trouble shooting technical difficulties when they arose. Site Coordinators were provided with the curriculum and instructor materials in advance of each class and were prepared to either co-facilitate instruction or deliver the instruction in the event of technical failure. Credit for both courses was available to students through Utah State University.

AREAS OF SPECIALIZATION

Generalist: At-risk and mild-moderate to severe disabilities

APPLICATION

Inservice instruction for paraprofessionals and classroom instructional teams

PROGRAM LEVEL

Inservice training:

TARGET AUDIENCE

Practicing regular and special education teachers and paraeducators working in special education and Title I programs

####### ONLINE COMPONENT

Courses were offered via a distance delivery system that provided live audio and video of the instructor and participants at multiple sites. Course materials were available to students and site coordinators on a project website.

TECHNOLOGY APPLICATIONS USED FOR ONLINE INSTRUCTION

####### COURSE MANAGEMENT SYSTEM/SOFTWARE AND HARDWARE

Project technicians developed a live, Internet-based delivery system with a high degree of technological control at the broadcast site and a low degree of complexity at the training site. Site coordinators at each reception site were required to identify and dedicate to project training a computer with at least a 90 MHz processor, 16 MB RAM, and Windows 95/98/2000/ME/XP. Personnel in participating school districts were required to identify classrooms or conference rooms as broadcast reception sites. Classrooms included typical instructional accoutrements along with the broadcast reception computer. Audio signals from the instructor and students at the various sites were heard over computer speakers and student audio signals from each site were delivered using Radio Shack unidirectional dynamic microphones. A Behringer MX802A Eurorack audio mixer was used at the broadcast site to manipulate sound quality from reception sites. At reception sites, instructional materials, the instructor's image and images of other students at broadcast locations were projected on television units or on to a wall or screen in the classroom using a LCD projector.

At the broadcast site, project staff delivered training broadcasts with a Dell Poweredge server and Canon XL1 mini-DV camcorder. Simultaneous live video and audio instruction were delivered using iVisit software. This program supported two-way audio and video and only required an Internet connection and three setting changes to operate. Ivisit software was available to the project as a free Internet download. To facilitate distance delivery of course content, project instructors required a slide tool such as Microsoft PowerPoint or Corel Presentation. Ivisit software did not support well known slide tool programs. However, iVisit software did support sharing of web pages. Therefore, PowerPoint slides were produced and converted to an individual website and then each website page was linked into a single website. Instructors delivered PowerPoint slides from the website, which when launched updated computers at each delivery site with the next slide.

Asynchronous Online Formats

Presentation of content. Asynchronous presentation of content with this project was minimal. Project technicians developed a Project IMPACT*NET website to post course materials. Site coordinators used the website to download session notes pages, readings, quizzes, and exams for students. These materials were provided to students before class began each week. Site coordinators and students were given separate login codes to access their individual sections of the website.

The student website listed students' scores on submitted assignments and provided access to project sponsored email.

Interactions with learners. At times, students and the instructors interacted using project-sponsored email. Interactions were predominantly questions posed to instructors concerning registration, assignments and assignment scores.

Synchronous Online Formats

Presentation of content. The ESP curriculum used with paraeducators consisted of five units including content on types of disabilities, federal legislation and mandates, the IEP process, the paraeducator's roles and responsibilities and their role in the IEP process, effective communication skills, how to relate to families of students with disabilities and persons of diverse ethnic and cultural backgrounds, behavioral assessment and intervention, student performance assessment (such as curriculum-based assessment), and assisting the teacher in delivering and monitoring instruction.

The CINC curriculum used with classroom teams consisted of six units including content on (a) establishing the teacher in the leadership role, (b) clarifying roles and responsibilities, (c) strengthening interpersonal communication, (d) responding proactively to management problems, (e) strengthening the teacher and paraprofessional in the classroom, and (f) evaluating staff performance. Both curriculum packages included instructor and student manuals, instructors' guides and VHS based video exercises displaying relevant school-based situations which required participants to analyze problems, make decisions and decide how they would respond in a similar situation. Content from both programs were adapted for online presentation. Key informational and discussion points were entered into PowerPoint slides, and video-based material was digitized using Windows Media technology. Once digitized, video segments were saved as individual files linked to a website and were streamed from the broadcast studio to reception sites for students to view. Students' primary learning activities included reading assigned materials, completing quizzes, listening to lectures, participating in site-based and across-site discussions, and completing course assignments and exams.

Interactions with learners. Synchronous interactions with learners occurred in three formats. First, students engaged instructors and students at other sites through the live audio/video system. Second, students interacted face-to-face with their local participants and site coordinator. Third, students engaged participants at other sites in cross-site discussions and activities. Finally, using the online system, site coordinators met frequently with project staff to review upcoming course materials, discuss problems and strengths of instructional delivery and review past, current and future student assignments.

FACTORS INFLUENCING DECISION TO USE ONLINE INSTRUCTION

NEEDS ADDRESSED BY ONLINE INSTRUCTION

Shortages of fully qualified special education teachers have been present in rural, suburban and urban America since the formal inception of special education. Public schools have mitigated special education teacher shortages by hiring paraprofessionals. Though hiring of paraprofessionals has been common practice, it is becoming more difficult to retain paraprofessionals. These factors are related to (a) the lack of pre-service training that teachers receive to effectively supervise paraprofessionals, (b) absence of basic job orientation and specific job training for paraprofessionals, (c) poorly defined job descriptions, (d) low wages, (e) limited opportunities for advancement, (f) perceived lack of administrative support, and (g) a reported lack of respect. These problems, combined with a dearth of local resources and identified training obstacles, prompted development of Project IMPACT*NET. Training obstacles included access to skilled trainers with knowledge and skills pertinent to paraprofessionals, access to college and university training programs, and awareness of and access to quality paraprofessional training curricula. The primary objectives of Project IMPACT*NET was, first, the development of a model for standardized, field-tested in-service training to enhance the instructional skills of paraeducators, and effectiveness of classroom teams and, second, development of a learning environment representing learners from a variety of districts and states.

Reasons for Selection of Specific Online Formats

Overcoming training obstacles and development of a diverse group of learners required the establishment of a distance education model which included a relatively inexpensive live, Internet-based delivery system which could be managed at broadcast and reception locations and could be easily duplicated by other organizations desiring to deliver instruction at a distant. This goal was accomplished using iVisit, a free download teleconferencing software, a broadcast-based PC with an Internet connection, a high quality digital video camera, and a server to stream video.

OUTCOMES AND IMPLICATIONS OF ONLINE INSTRUCTION

Impact of Online Instruction on This Program

A primary outcome was, over a four-semester period, 133 paraeducators and 68 members of classroom teams participated in training to strengthen paraeducators' skills of and the skills of instructional teams. Soon after training began, the No Child Left Behind (NCLB) mandate required, among other things, that teachers and paraeducators to be identified as "highly qualified." According to the mandate, "highly qualified" could be demonstrated in multiple ways, including completion of training. When the project began, participating school districts were not providing inservice training to paraeducators or to classroom teams, and therefore Project Impact*NET provided teachers and paraeducators in four states with inservice training that they would not otherwise have received.

Opportunities/Advantages/Benefits of Online Instruction

Connecting paraeducators and supervisors across the U.S. in a single live classroom environment is the most apparent benefit of this project. Teachers and paraeducators in multiple locations in the U.S. were able to see and hear one another and engage in site-based and cross-site discussions. This forum set a context for teachers and paraeducators to identify common student and job related problems and learn of the rich variety of approaches being applied to solve problems. Cross-site discussions and in-class assignments were fortuitously timed with passage of NCLB. This cross-site engagement allowed paraeducators and teachers to share their fears and frustrations associated with the mandate and approaches being taken by their respective districts and State Departments of Education in response to "highly qualified" requirements, highs takes testing, and other provisions of the act. Comments provided by participants best describe the value of technology when it is effectively coordinated to provide quality instruction and discussion at the local sites, distant sites, and across sites:

"This class provided a great opportunity to share and compare ideas used by other paras that we would not have the opportunity to learn from."

"Being able to participate in this class as a team has allowed us to immediately apply the things we have learned in a meaningful and relevant manner."

"I learned a lot of practical, useful techniques. It was very informative and really applied to my job. The video examples were good."

Problems/Disadvantages/Limitations of Online Instruction

The advantages of online instruction were accompanied by a variety of problems or obstacles including the following:

1. It was difficult to identify a satisfactory time to hold class when participants resided in separate time zones, creating a 2-hr divide;

2. It was difficult to fully anticipate the breadth or depth of the role site coordinators play at their distant locations. Over time, site coordinator roles and responsibilities took the shape of teaching assistants versus technical aides or course management aides, e.g., registering students, administering quizzes and tests and providing copies of instructional handout.

3. Technical difficulties at the delivery site included unannounced closure of the university Internet hub prior to class and insufficient university bandwidth due to unannounced events consuming bandwidth required for course delivery.

4. Examples of technical difficulties at the reception site included installation of updated firewalls resulting in closure of portals used for reception of class broadcasts.

5. Closure of firewall portals after class to maintain the integrity of the firewall and then failure to open them for class. Again, participants' comments illustrate some of the challenges associated with online delivery:

    "It was hard to understand what other sites were saying. The audio was delayed and either too loud or too soft."

    "We had problems at times, but it was always immediately remedied."

"I wish we could have seen them (i.e., other sites) better."

Suggestions or Recommendations for Others Considering Online Instruction

There are a host of suggestions that we could offer and many of them would be useful. However, the following suggestions represent some of our most important learning. First, in advance of course delivery, develop memorandums of understanding (MOU) with technical coordinators in participating school districts. The MOU should specify the availability of firewall portals well in advance of each class session and the need for technical coordinators to notify instructors of installation of firewall updates. In addition, technical coordinators should arrange "dry runs" of the delivery system following updates. Second, discuss with university technical coordinators the technology being used and the days and times of delivery and, together, analyze the historical bandwidth load on class days and at instructional times. Formalize an agreement that technical coordinators will contact instructors well in advance of updates or other alterations or events that may impair instructional delivery. Third, select site coordinators with technical skills, effective communication and instructional skills, and demonstrated support of key district personnel (e.g., Superintendent, Special Education Director or the Technical Coordinator). Fourth, define the role of site coordinators in terms of their technical, course management, and general instructional responsibilities. In addition, clearly define the site coordinators specific role as "instructional assistant" to the instructor. Finally, clarify for site coordinators their responsibility to act as a liaison between instructors and key district personnel.

LIST OF RESOURCES

Curriculum

Colleagues in the Classroom (CINC)
Enhancing the Skills of Paraeducators (ESP)

Contact Information

David E. Forbush, PhD
Assistant Professor
Utah State University
Department of Special Education & Rehabilitation Counseling
2865 Old Main Hill, Logan Utah
84322-2865
(435) 797-0697
davidf@cc.usu.edu

Robert L. Morgan, PhD
Associate Professor
Department of Special Education & Rehabilitation Counseling
Utah State University
2865 Old Main Hill, Logan Utah
84322-2865
(435) 797- 3251
email: bmorgan@cc.usu.edu

Chapter 12

The Professional Development School without Walls in Southeast Florida

Betty C. Epanchin
Elizabeth Doone
Karen Collucci
University of South Florida

Title of Program/Sponsor

Professional Development School without Walls
College of Education at University of South Florida

Brief Description of Program

Overview of Program. The Professional Development School without Walls is a partnership between the Department of Special Education, College of Education at the University of South Florida (USF) and the two large Florida school districts located geographically near the university, Hillsborough and Pasco Counties. The partnership was developed for the purpose of improving practice both at the university and in the school districts, and, over the years, a number of mutually shared problems were addressed. This program description addresses one of the earliest partnership projects.

At the time the project was conceived, the special education teacher preparation program had over 120 candidates entering the program annually, which created challenges in staffing the program both at the university and in the schools. The teacher education faculty at the university could not supervise such large numbers of teacher candidates using the traditional model of supervision, and sufficient numbers of adjunct faculty who could provide quality supervision of the teacher candidates' clinical experiences were unavailable. Even when competent adjuncts were available, they rarely had the time to be closely affiliated with the university program; thus a disconnect was created between the on-campus teacher preparation program and the clinical experiences.

The schools also had difficulty identifying sufficient numbers of motivated, competent cooperating teachers to supervise the teacher candidates. While there were a number of teachers who had the potential of being outstanding mentors, these teachers were called upon by their schools to carry out a number of leadership roles within their schools. Feeling overburdened already, many were reluctant to assume the responsibilities of supervising a teacher candidate.

A further challenge resulted from the policy one of the districts adopted regarding the placement of teacher candidates. Because of the large numbers of students the district was placing (almost 1,000 students annually when all education areas were considered), one of the school districts decided they could no longer collaborate regarding placements. The district staff development officer worked with school principals in identifying possible placements and assigned teacher candidates to schools without input or consultation from the university programs. Too frequently, this resulted in placements that did not match the university program's teaching philosophy. Teacher candidates were not seeing classroom practice that matched the content of their teacher education curriculum, which often resulted in their feeling discouraged and disillusioned about teaching. In spite of the problems this practice created, the university clearly understood the challenge faced by the district when trying to find placements for large numbers of students each semester.

To address these challenges, the Professional Development School without Walls partnership created a new teacher preparation role which focused on identifying, preparing, and supporting quality mentors in the schools who would be responsible for providing the supervision of teacher candidates. Rather than placing candidates in professional development schools, which would limit the number of schools that could be potentially involved, master teachers who wanted to partner with the university in the preparation of special education teachers were identified. In creating this role, careful attention was given to identifying rewards and incentives for the teachers so that supervision of a teacher candidate would not become just another set of demands. Teachers were selected from schools throughout the two counties, thus enabling a broad base of school participation. Teacher candidates were placed with these mentors, called Professional Practice

Partners (PPPs) for their final internship.

To qualify as a PPP, teachers were required to submit written applications that were screened by a representative from both the district and the university. The screening consisted of a review of their paper application to determine that the teacher was tenured, recommended by her principal, and had successful experience functioning in some type of mentoring role. Teachers who passed the screening were then interviewed, again by a team from the university and the school district. Interviews consisted of approximately 10 questions asked of all candidates. These questions dealt with their perceptions of the mentoring role and their educational philosophies.

Once a teacher was selected, she was required to participate in a semester-long graduate seminar about supervision and mentoring. The interviews and seminar discussions were both designed to ensure that PPPs were committed to promoting growth within each candidate, rather than replicating their own classroom practice.

Once a teacher completed the training and was identified as a Professional Practice Partner, he or she functioned as both cooperating teacher and university supervisor for the teacher candidates, mentoring and guiding students' work during their student teaching semester. To support their work as supervisors, PPPs participated in ongoing professional development seminars led by university faculty designed to provide them with support and opportunities for continued growth. These seminars also created a forum for sharing observations and information about the teacher candidates so that, if a problem arose with one of the candidates, a university faculty member would become involved in resolving the problem. No PPP was ever expected to deal with a challenging supervisory experience alone.

Area(s) of Specialization

Special Education Field Experiences

Application

Senior year of undergraduate program

Program Level

Preservice personnel preparation

Target Audience

Senior undergraduate teacher candidates and their mentors

Online Component

Pro-seminar electronic support offered by mentor during student teaching field experience

TECHNOLOGY APPLICATIONS USED FOR ONLINE SUPPORT AND TRAINING

Course Management System/Software and Hardware

Concomitant with the support and training provided to the PPPs, university faculty also provided teacher candidates with primarily an electronic support structure. From the beginning of their programs, teacher candidates were assigned to small seminar groups, called Professional Seminars (Pro-Sems) that were designed to promote the integration of theory and practice and to assist students in developing their own personal teaching philosophies and beliefs. Groups of 8 to 10 students met weekly with their university faculty advisor for discussions about themselves as ethical and caring teachers. The Pro-Seminar leaders were responsible for the same group of 8 – 10 students for the duration of their time in the program to provide a continuity of care. In such a large institution as USF, this was intended to foster personal relationships between students and with faculty. During the first three semesters of the teacher preparation program, Pro-Sem facilitators met with teacher candidates as part of their scheduled classes, but, for the final field experience, the meetings were conducted for the most part electronically.

The final field experience, student teaching, traditionally had a bi-weekly Senior Seminar attached to student teaching. With the advent of PDS without Walls, the Senior Seminar was shifted primarily to an electronic format. All candidates in student teaching were instructed in the use of Blackboard (most were familiar with it) and assigned a password and user id prior to the first week of their final internship. The syllabus was posted to Blackboard, as were the weekly assignments. Two of the assignments were particularly well suited to the blackboard format and especially effective in facilitating communication, growth and learning.

Asynchronous Online Formats

Presentation of content. The critical incident was a weekly assignment that involved the entire group of 8 to 10 candidates. Each week one or two of the group members were assigned the responsibility of posting a critical incident. Candidates who were not assigned the responsibility of posting an incident were required to respond to the incident that was posted. The critical incident assignment is shown below.

Critical incident report

Consider all the interactions and experiences that you are having. Select one that has caused you concern and challenged your notion of knowing what to do. This should be an incident that occurred at your school or in your classroom and one that you felt involved some ethical issues with no obviously right solution. Describe this incident, which we shall call the "critical incident." providing the following details:

- What happened?
- Who was involved?
- How you responded to the situation and why you responded in this fashion?
- Did your response resolve the situation, temporarily delay the situation, or do you still need advice and closure?
- How did you present closure to those involved?
- What were your cooperating teacher's/ PPP's thoughts about the incident?

These incidents will be posted in the discussion board section of Blackboard. Incidents must be posted by 5:00 p.m. on Wednesday.

Respondents to the incident should post their impressions and discuss what might have been done differently. Responses to the incidents must be posted by the following Wednesday at 8:00 a.m.

All students were encouraged to submit additional critical incidents as desired. Students responded positively to this format, describing a wide variety of challenging situations and asking for help. For example, one candidate felt the paraprofessional in her classroom undermined her authority. She did not know how to approach her mentor teacher about this because she thought the teacher and the paraprofessional were personal friends. She was able to solicit good advice from her classmates by posting the questions. Her classmates, who were not so emotionally involved, were able to provide her with helpful insights and suggestions. The process also reminded student teachers of the value in seeking multiple viewpoints. Additionally, by reading a descriptive account of a critical incident in one of their classmates' classrooms or schools and by being involved in the problem-solving process and follow-up, student teachers were able to virtually experience more than what was happening in their own classroom and learn from the experiences of their classmates. When asked to evaluate this assignment and format, university facilitators noted that candidates were not as reserved when discussing sensitive issues as they typically were in person.

Another of the more effective Blackboard assignments was the requirement that students submit personal reflections weekly, as detailed below.

> **Reflection**
>
> Reflection is a critical component of becoming a better teacher. To promote your use of reflection, you are required to complete electronic journals twice per week. These entries should briefly describe your reactions to the day's events. Journals should include the following information:
> - brief description of a lesson you taught, a situation you managed, or an extended interaction in which you were involved (the event)
> - brief description of your feelings and thoughts about the event
> - an evaluation of your action, including both your successes and ideas about how you might handle a similar event differently in the future
> - thoughts or plans for the next lesson or event
>
> Journals will be e-mailed as attachments through the communications section of Blackboard (see info below for access instructions) to your Senior Seminar Facilitator and are due on Wednesday and Friday by midnight. Your Senior Seminar Facilitator will provide electronic feedback weekly.

In evaluating this assignment and the on-line format, the most frequent response from facilitators had to do with their positive feelings about having everyone respond rather than the most vocal. As one facilitator wrote, "I had student teachers who did not talk in class. I was not sure if they were just more reserved or tired from their day of student teaching, but they would write pages in the weekly journal. These students often sent more than the required journal entries." Both student teachers and the facilitators commented on the fact that the electronic format enabled them to provide immediate feedback to the student teacher, as well as a format for continuous follow-up. The dialog thread could be continually revisited as additional methods were tried, or circumstances evolved. The format also provided a real sense that problems are not just solved, but rather they require on-going fine-tuning and repeated revisiting, which provided students with a more realistic picture of classroom issues. Facilitators also reported that the format was more convenient for student teachers with very busy schedules. As one facilitator said, "I think the online format is good because they are able to complete their assignments at their convenience and it's easy to recall sent/deleted emails if something was missing."

Interactions with learners. To ensure that the PPPs were kept in the communication loop regarding student teachers' progress, monthly meetings with them were scheduled during the semester in which they supervised/mentored a student teacher. As the electronic communication system developed, the PPPs were added to the weekly postings. Faculty could use the posting to generate questions to the PPPs for discussion at the monthly meetings. Additionally, the postings were used to facilitate problem-solving sessions among the PPPs. Over time, the PPPs began utilizing a "problem- solving" mode rather than a "things aren't going well" mode when challenges arose with their student teacher. During the monthly meetings the PPPs also developed a shared sense of problem solving and recognized the value of learning from and supporting each other. For example, an inexperienced PPP shared her feelings of inadequacy at one of the meetings. She felt she was a less than capable mentor because she was not able to pick up on her student teacher's undermining and deceptive behavior around professional issues. The other PPPs were able to share similar stories from past mentoring experiences, and then they helped her problem solve ways to provide him with feedback and guidance, and develop a plan to successfully complete the semester. By the end of the semester, the PPP felt she had grown tremendously and she attributed her growth to the support of the other PPPs. The collaborative work of the teachers truly created a learning community that spanned across many schools and school districts.

To further enhance and support the work of the PPPs, an online newsletter was created to provide updates, general information and feature research initiatives. Highlights have included:

- Reports from collaborative research groups
- Lists of PPPs who recently passed the National Boards
- Birth Announcements/ adoption announcements
- Upcoming workshops information
- Book club reading list
- Training dates and times
- Listserv address and how to contact PPPs

Although the newsletter was positively reviewed by the PPPs, they were not able to assume responsibility for its production due to the varied additional responsibilities they already perform at their schools.

FACTORS INFLUENCING DECISION TO USE ONLINE SUPPORT

Reasons for Selection of Specific Online Formats

Technology has the capacity to enhance both instruction and field supervision in teacher preparation programs as well as to facilitate communication between schools and universities – a vital feature of quality partnerships. The online format bridges the extensive distances teacher candidates and supervisors must travel for face-to-face meetings. Teacher candidates and supervisors have easy access to frequent and easy communication, which builds trust and creates a community of learners. Teacher candidates are able to complete work and reflect on their practice at times most convenient to them, which also relieves stress.

OUTCOMES AND IMPLICATIONS OF ONLINE SUPPORT

Impact of Online Instruction on This Program

Feedback from both the student teachers and the PPPs has been very positive about the on-line component of the senior seminar and the use of technology to further communication.

Opportunities/Advantages/Benefits of Online Support

Technology allows for more timely communication and feedback. Oftentimes, the new roles and responsibilities of the teacher candidate place them in challenging situations about which they want and need timely feedback and encouragement of his/her peers and senior seminar facilitator. Also, we have found that the online format breaks down inhibitions to speak candidly about unethical situations, areas of concerns, personal conflicts and other issues that arise in schools that often are difficult for teacher candidates to discuss.

Suggestions or Recommendations for Others Considering Online Instruction

Technology is a tool that enhances the experiences in teacher education programs, facilitates the mentoring process, and creates a supportive community of educators focused on continuing to support the developmental growth of new teachers.

CONTACT INFORMATION

Dr. Betty Epanchin
Professor and Associate Dean for Teacher Education and School Relationships
Director, The Teachers Academy
918 Curry Building
P.O. Box 26170
Greensboro, NC 27402-6170
Office: 336-334-3949
Fax: 336-334-4120
bcepanch@uncg.ledu

Dr.Elizabeth Doone
Visiting Assistant Professor
Instructor and MAT Program Coordinator
Department of Special Education

EDU 162
College of Education
University of South Florida
4202 E. Fowler Ave.
Tampa, FL 33620
813-974-9929
edoone@tempest.coedu.usf.edu

Dr. Karen Colucci
Visiting Assistant Professor
Instructor and MAT Program Coordinaotr
Department of Special Education
EDU 162
College of Education
University of South Florida
4202 E. Fowler Ave.
Tampa, FL 33620
813-974-1398
colucci@tempest.coedu.usf.edu

Chapter 13

SUPPORTING MASTER OF SCIENCE IN SPECIAL EDUCATION STUDENTS ONLINE: CHALLENGES AND TENTATIVE SOLUTIONS

JOAN P. SEBASTIAN
STUART SCHWARTZ
JANE DUCKETT
National University

TITLE OF PROGRAM/SPONSOR

Master of Science in Special Education Program
School of Education at National University, La Jolla, California

BRIEF DESCRIPTION OF PROGRAM

Overview of Program. The Master of Science in Special Education is a School of Education graduate level program that requires core courses followed by advanced specialization courses. Courses are offered both onsite and online. This program of study is designed for educators and other professionals who want to become knowledgeable about educational learning problems and teaching strategies to enhance student performance.

The program is based on the premise that meeting the special instructional needs of students in today's schools requires knowledge of a wide array of teaching strategies since no one strategy can meet the needs of every student. Another major premise is that all prospective teachers must be aware and accepting of cultural, linguistic, ethnic, racial, economic, gender, sexual orientation and ability differences, and they must achieve mastery of methods and techniques that will accommodate the increasing diversity in contemporary special education and general education programs where special needs students are served. The program is designed to present a variety of research-validated methods, techniques, and approaches that will empower the prospective teacher to weave together instructional programs that will positively impact the special needs of K-12 students across every developmental domain. Faculty members acknowledge and value the need for a full range of service delivery options for serving special needs students. The program has the following learning outcomes.

By the end of this program, students will be able to:

- Apply research methods including critiquing and synthesizing current educational literature.

- Utilize a variety of methods and technology applications in order to complete action research and thesis projects.

- Demonstrate expertise in the use of technology for instruction as evidenced by the development of adaptive devices, videos, toys /manipulatives, and software.

- Summarize federal and state laws, policies, and major court cases regarding exceptional individuals.
- Design and implement positive behavior supports for students with special needs.
- Utilize skills in instructional design including adaptations required to assist special education learners to achieve state standards-based curriculum.
- Demonstrate consultation and collaboration skills, including the ability to implement professional development at sites /districts for service delivery across all curricular areas and disabilities.
- Locate appropriate resources and practices in advocating for exceptional individuals.

Area(s) of Specialization

Mild/Moderate Disabilities
Moderate/Severe Disabilities

Application

Post-baccalaureate Level 1 Education Specialist Credential and/or Master's degree program

Program Level

Preservice personnel preparation

Target Audience

Non-traditional adult learners who may also be unlicensed educators working in special education settings throughout the State of California and beyond

Online Component

All courses, excluding the assessment course, are offered both online and at various campus locations in California and Nevada. Field study courses are offered at the local site with locally prepared supervisors or adjunct faculty. Learners come to regional Learning Centers for demonstration of specific assignments and competencies related to the Field components when possible. If distances are too great, competency demonstration is facilitated at the local, rural site. All final supervised teaching (student teaching) occurs at the local site in candidates' assigned classrooms.

TECHNOLOGY APPLICATIONS USED FOR ONLINE INSTRUCTION

Course Management System/Software and Hardware

The main delivery system for online instruction is Blackboard, one of several commercial programs available to support a networked, online learning model. Blackboard has been used by National University for the past two years. Earlier online programming utilized eCollege. The switch to Blackboard necessitated the revision and updating of all online courses and extensive training of faculty and online support staff. Prior to students enrolling in online courses, they are expected to complete an initial Blackboard training program, which is done online, so that they are prepared to navigate through the system. Constant support is available for students and faculty so that questions or problems can be addressed as needed. The support is provided, via email and by toll free phone, seven days a week by staff members of Spectrum Pacific Learning, which is a branch of the National University System

Asynchronous Online Formats

Presentation of content. Content in each of the courses is typically formatted in either 4 or 8 units of instruction. Units are built around course themes, which respond to the California Standards for the Preparation of Teachers as specified by the California Commission on Teacher Credentialing. Most courses are offered over a 4-week/1-month time frame. Selected courses are available in a 2-month format due to course requirements. Course content includes (a) readings online as well as text or print based material, (b) multi-media slides/video and audio embedded in the course and on CD, (c) case studies, (d) assessments, and (e) assignments to be applied in field settings. Students have access to the National University online library resources.

Interactions with learners. There is a high level of interaction among students and faculty. This is accomplished by having mandatory participation in asynchronous discussion boards for each unit of instruction. Typically the professor posts 3-4 questions within each unit and students are required to become actively engaged in the discussions. Directions given to students explain that substantial contributions to the discussions are expected and that credit is not given for such remarks as, "that's interesting," or "good that you posted that." Informal observations made by students and faculty involved in the online courses suggest that students have a higher level of participation in online, rather than in onsite, courses. This appears to be due to the requirements for postings in online discussions; students often are able to sit quietly in onsite courses with only a minimal amount of interactions and very few contributions.

In addition to the asynchronous discussion board assignments, students interact with their peers and instructor via email. In some classes, Blackboard discussion groups are established for this purpose and specific assignments are given to the groups. Instead of using the Blackboard groupings, some instructors simply suggest email for student group interactions and for direct communication with students.

Group emails and listservs are used in some courses to supplement the announcements section within Blackboard. These communication methods allow instructors to either send information to all students outside of the Blackboard system or, if preferred, enable class discussions with participation from everyone via email. While easy to use, emails and listservs are external to the Blackboard system and therefore no automatic record is kept of student participation or of the actual discussions.

SYNCHRONOUS ONLINE FORMATS

Presentation of content. Currently, only two courses make regular use of the live online chat capabilities of the Blackboard system. The first course, EXC 650 Consultation and Collaboration, requires that students work in small groups to role-play, in real time, simulated situations that require application of skills described in the course content. Real time interactions require a considerable amount of coordination on the part of the instructor. Because many students work in schools during the day and also take classes on campus two evenings during the week, scheduling evening discussions is a challenge. Multiple times for groups to meet are typically offered so that all students can find a time that meets their schedule in order to complete the assigned activities.

In EXC 657 Community Resources and Transition, regularly scheduled synchronous chats are held during the course. These are used as optional drop-in sessions so that students can ask questions and participate in discussions with other students and the instructor on both course content and assignments. Students who have participated in fast-paced discussions in other online environments (such as AOL chat rooms) appear to be very comfortable with these chats; those who have never been involved before with similar activities seem to have difficulty becoming engaged in the conversation and contributing appropriately.

Another real time interaction that occurs in the program is during the specialization courses. Each 4.5 quarter credit specialization or methods course also has a 1.5 quarter credit field experience attached to it that requires students to come to campus and demonstrate skills and interact with classmates. When one of the specialization courses is offered online, the corresponding field studies course meets twice during the month at Learning Centers located throughout the state. Students living at very remote sites and unable to travel to a Learning Center are able to submit assignments and demonstrate competencies to local facilitators for evaluation. Local facilitators receive training and supervision from full time NU special education faculty.

National University recently subscribed to the iLinc system, which is expected to greatly enhance opportunities for synchronous classes or class components via the Internet. iLinc enables an instructor to meet with students while they are at their own computers using an audio and video connection. The students can hear their instructor and other students, respond in writing or by voice, view a dynamic whiteboard and add to it, and see and hear a variety of teacher-presented materials, such as video clips and power point slides with accompanying audio clips. This new system has yet to be used on a large scale; several faculty members are currently experimenting with its use for both classes and meetings. It is expected that entire courses will be taught synchronously with students from around the globe with this new technology.

Interactions with learners. Students respond in small groups to online discussion groups/chats that are in required as part of a course. Additionally, students come together to meet in field studies classes twice a month at Learning Centers to share information and course expectations. In the online experience, students typically respond to case study problems and work in small groups to identify strategies for resolution. Students also are able to discuss course issues with the instructor in a live chat situation.

Online students completing the final research project in the program (either a thesis or action research study) also have multiple opportunities to interact in real time with the course instructor at a local Learning Center or over the telephone. Instructors have reported that direct real time contact, whether by phone or in person, has helped to support students in the development of their research proposal. One-on-one support allows for personalization of each student's classroom based research and facilitates project completion.

FACTORS INFLUENCING DECISIONS TO USE ONLINE INSTRUCTION

Needs Addressed by Online Instruction

National University was established in 1971 with the mission of serving the unique needs of adult learners throughout the State of California. The one course per month format, Learning Centers located throughout the state, and full programs offered online allow graduate teacher candidates maximum flexibility to complete credential programs in special education. A recent survey of online special education students indicated that over 50% take courses because of the flexible schedule that allows for meeting family and work responsibilities while completing credential requirements (Sebastian, Caywood, Duckett, & Swenk, 2003). Online learning provides access for students in rural communities throughout the state. Interestingly, urban students report that completing classes online saves them the time and "hassle" of commuting on overcrowded highways in the cities.

California, as other states throughout the country, experiences a critical shortage of fully qualified special education teachers. This is true in both urban and rural school districts. Eighty percent of current students completing courses in special education at National University are working in special education classrooms on emergency or internship certificates.

Reasons for Selection of Specific Online Formats

National University decided to implement the Blackboard course management system for several reasons. Blackboard allowed National University to develop its own internal online support structure, which is currently operated by Spectrum Pacific Learning, an entity that is part of the National University system. The online support, which is internal to National University, can be more responsive to faculty and student needs. Blackboard also provided extensions and additional upgrades for the initial system in a timely and efficient manner.

OUTCOMES AND IMPLICATIONS OF ONLINE INSTRUCTION

Impact of Online Instruction on the This Program

In 2000, National University began offering special education courses online using the e-College course management system. Courses in the Generic Core sequence of the program, which are required for all credential and MS students, were the first to be developed for online delivery. Instructors who taught the course on campus were asked to develop the courses online, with the assistance of technical help from e-College. The assessment course was the only course not developed for online delivery due to the nature of the content. Course developers felt that teaching students assessment skills with appropriate test instruments needed to occur on campus where students could have access to test instruments and receive immediate instructor feedback on performance.

In 2002, the remaining courses in the credential and MS program were developed online using the Blackboard course management system. The only courses not currently offered online remain the assessment course and the specialization field studies courses, as well as the final student teaching experience. Most recently, students completing a Master of Science degree have been able to complete their final research project online as well.

Opportunities/Advantages/Benefits of Online Instruction

Online instruction provides learning opportunities for many students who would be unable to complete a credential or Master's degree program. Many of these students live great distances from university programs, or have family and work obligations that affect their ability to participate.

During the past 4 years of operation the online program has grown rapidly. Currently, over 25% or 369 special education students complete at least one course online. Additionally, over 400 current students will have completed more than half of their program online. Students report satisfaction with most aspects of the online program and believe they have gained important knowledge and skills to serve students with disabilities (Caywood, Sebastian & Duckett, 2004) An analysis of student teaching competency evaluations suggest that there is no significant difference between the performance of students who have completed key specialization courses online vs. on campus students (Duckett & Shaddock, 2004).

Problems/Disadvantages/Limitations of Online Instruction

Updating and maintaining online courses is an ongoing effort that requires attention. While support services for all online courses are provided by Spectrum Pacific Learning for faculty and students, the need to update course content remains the responsibility of faculty in the special education department. Finding time to update courses along with ongoing teaching, service, and scholarly responsibilities has been a challenge for all full time faculty.

A recent degree requirement, instituted in the fall of 2003, for a full thesis at the end of the MS program in Special Education appears to have impacted enrollment adversely. Students view credential courses and field activities as relevant for their roles as teachers but are fearful of or do not see the value of completing a research project or thesis. Faculty report that students are struggling with the rigorous requirements of a thesis. Specifically, they are not well prepared to complete the extensive literature review, develop appropriate research methodology and gain Institutional Review Board approval for the study. Very few students have completed the thesis; many more are struggling at various stages of completion.

In response to these issues, the faculty has implemented an Action Research option for the final course in the MS program. Action research requires that the students systematically study their own teaching practice. Students completing an Action Research project are required to design a study that involves the basic research cycle, beginning with problem identification, review of relevant literature, development of study questions, data collection and analysis, and concluding with a discussion of implications. While this is similar to thesis requirements, Action Research studies are limited to the graduate student's own classroom. Students also are required to present their Action Research study to colleagues, either in a National University course or with the faculty at their school. NU faculty take the position that preparing students to study their own practice, in a systematic and rigorous manner, is a more relevant and practical culminating activity for the MS degree. Students also report that the Action Research project is more useful than completing a more traditional thesis. As of the Winter 2005 term, MS in students in special education will be required to complete an Action Research study as the final program requirement and only those students whose career goals require more expertise in statistics, evaluation, and research will be encouraged to complete a thesis.

SUGGESTIONS OR RECOMMENDATIONS FOR OTHERS CONSIDERING ONLINE INSTRUCTION

Designate a special education faculty member to be the

online coordinator for your department. It should be this person's responsibility to be sure that instructors have the skills needed for teaching online. This person should monitor courses to ensure that faculty members are providing instruction as expected. Sufficient time must be allotted so this faculty member is able to properly handle these responsibilities.

While some software programs support editing of papers, such that the students see instructor remarks and suggestions, it seems that it's easier and more efficient for papers to be submitted either within a program, such as Blackboard or as documents via email. Then, the instructor is able to read, remark, and then discuss the paper with the students either via email or telephone conversations.

Wireless laptop computers should be made available for all faculty members who are teaching online. This enables them to access their courses on a daily basis from work, home, or while traveling. Travel budgets need to allow for Internet access fees until the hotel and extended travel industry (for instance train stations and airports) include free wireless access without charge.

LIST OF RESOURCES

BOOKS AND ARTICLES

Gay, L. R., & Airasian, P. (2003). *Educational research competencies for analysis and applications.* 7th Ed. Upper Saddle River, NJ: Pearson Education, Inc.

Mills, G.E. (2003). *Action research a guide for the teacher research.* 2nd Ed. Upper Saddle River, NJ: Pearson Education, Inc.

Sebastian, J. P., Duckett, J., Caywood, K., & Swenk, J. (2003).

Becoming a special education teacher online: Candidates perceptions. In R. Menlove (Ed). *Conference Proceedings: American Council on Rural Special Education (ACRES).* March 19-22, 2003. Salt Lake City, UT.

Duckett, J. & Shaddock, G. (2004). *Evaluation of student teaching program.* Presentation at the Teacher Education and Special Education (TTED) Conference, November 10-13, 2004. Albuquerque, NM.

Web Sites

http://www.nu.edu
The National University web site and portal for all online courses.

http://www.ilinc.com
The iLinc website which describes the technology for audio and visual Internet interactions.

http://www.scu.edu.au/schools/gcm/ar/arhome.html
The home page for a host of action research-related resources. There are links to sites ranging from a simple definition of action research to electronic journals, organizations, and papers.

Contact Information

Joan P. Sebastian, Ed.D.
Department of Special Education
School of Education
11255 North Torrey Pines Road
La Jolla, CA 92037
jsebasti@nu.edu

Chapter 14

ONLINE MODULES FOR DISTANCE EDUCATORS AT THE UNIVERSITY OF KENTUCKY

BELVA C. COLLINS
CONSTANCE M. BAIRD
University of Kentucky

TITLE OF PROGRAM/SPONSOR

Distance Education Certificate Courses
College of Education at the University of Kentucky

BRIEF DESCRIPTION OF PROGRAM

Overview of Program. The University of Kentucky (UK) has had three decades of experience in distance education that include the use of satellite, interactive video, Web-based technologies, and video-desktop conferencing. In particular, the Department of Special Education and Rehabilitation Counseling (EDSRC) at UK has operated a graduate distance education program since 1989. In 1996, the EDSRC received a federal grant from the U.S. Department of Education to develop a doctoral program with a focus on leadership in distance education. As part of the doctoral program, a faculty member with experience in the use of distance education technologies developed a course in distance education delivery that was taught on-campus to the EDSRC doctoral students. Based on the success of this course, the faculty member and an administrator from UK's Distance Learning Programs developed two subsequent Web-based distance education courses to prepare distance educators. The course development process included (a) meeting with a focus group of experienced distance educators across disciplines at UK and an affiliated community college to identify course topics, (b) gaining approval from UK administrators, (c) selecting a delivery mode, (d) conducting an extensive review of the distance education literature, (e) securing course development funds from UK, (f) identifying course development team, (g) developing timelines, (h) interviewing experienced distance educators, (i) editing video clips of distance education delivery techniques, and (j) preparing graphics. The delivery of the courses began in the fall of 2000. Each is taught on an annual basis and continues to evolve in both content and delivery mode.

The first course, Distance Education: Delivery, is taught by the special education faculty member and covers the following topics: (a) an overview of distance education technology, (b) planning a distance education course, (c) preparing materials for distance education delivery, (d) planning for interactions in a distance education course, (e) planning a distance education class session, (f) testing and evaluation, (g) preparing students and faculty as distance education participants, (h) considering undergraduate and graduate issues in distance education, (i) comparing distance education models, and (j) working collaboratively to deliver distance education coursework. The second course, Distance Education: Management and Support, is taught by the distance education administrator and covers the following topics: (a) historical overview of distance education; (b) distance education technology systems; (c) planning and development of distance education programming; (d) academic, administrative, and technological support services; (e) distance education library/electronic resources and virtual libraries; (f) instructional funding of distance education; (g) distance education policy issues; (h) distance education structure within higher education; (i) institutional collaboration; (j) emergence of new distance education models; and (k) the changing environment and future direction of distance education. Both courses are part of a distance education certificate currently being developed in the UK College of Education in collaboration with the Department of Curriculum and Instruction, which will offer courses in instructional design.

AREAS OF SPECIALIZATION

Distance Education Delivery
Distance Management and Support

APPLICATION

3 hrs of graduate credit per course
May be used as a doctoral area of specialization or as professional development

PROGRAM LEVEL

Graduate or Post-graduate Studies

TARGET AUDIENCE

The course is open to distance educators and future distance educators who are faculty or graduate students across disciplines both within and outside of UK. UK has an agreement that the courses can be offered outside of Kentucky at an in-state tuition rate. The majority of students have been doctoral students in special education from UK or Utah State University, graduate students across disciplines from UK, or faculty members across disciplines from UK and the Kentucky Community and Technical College System.

ONLINE COMPONENT

The courses are now totally online following an evolution from traditional on-campus delivery to a hybrid of on-campus and Web-based delivery. Occasional optional on-campus meetings are held at student request with audio-conferencing used to include distant students in these meetings.

TECHNOLOGY APPLICATIONS USED FOR ONLINE INSTRUCTION

COURSE MANAGEMENT SYSTEM/SOFTWARE AND HARDWARE

The courses currently utilize a Blackboard course management system after migrating from TopClass when UK changed systems. Course components consist of videoclips, audioclips, and PowerPoint slides. The courses deliver content through three types of video- or audio-clips: (a) videoclips of interviews with experienced distance educators, (b) videoclips of distance education techniques from special education courses taught through satellite and interactive video delivery, and (c) video- or audio-clips of the course instructor presenting content on each course topic. To access videostreamed content, students must download RealPlayer from the Internet.

The two instructors created the interview videoclips by editing videotape of interviews they conducted with experienced distance educators in the UK television studio, at distance education delivery sites, or through submitting written questions that distance educators videotaped at a distant site and then mailed to the instructors. Those interviewed represented the UK campus, Lexington Community College, the Kentucky Community and Technical College System, regional institutions within Kentucky, universities in other states, and distance education consortia. In addition, they reflected distance education experiences in agriculture, biology, communication disorders, communications, deafblind intervention, early childhood, education administration, engineering, family studies, library and information science, music, physical therapy, rehabilitation counseling, special education, technology and media design, and virtual universities.

The faculty member created videoclips of distance education techniques by editing videotape from four courses previously offered through the EDSRC at UK. From the first course (Basic Skill Training for Students with Severe Disabilities) in which she was the primary course instructor, she selected videoclips that demonstrated techniques for course delivery to a singe site that used both satellite and interactive video across class sessions (e.g., delivery of lecture content, use of graphics, facilitation of interaction between on-campus and distant students), and interaction with a team teacher. From the second course (Transdisciplinary Services for Students with Multiple Disabilities), she selected videoclips that showed interactions between team teachers and students across universities within the state using interactive video technology. From the third course (Applied Behavior Analysis), she selected videoclips that showed a noted faculty member from another university in another state presenting a guest lecture and interacting with students using interactive video technology. From the fourth course (Distance Education: Delivery), she showed videoclips of doctoral students practicing the use of interactive video and satellite technology during simulated distance education class sessions.

To deliver course content, the administrator chose to pretape audioclips in the UK television studio that later were synchronized with PowerPoint slides of key points. To deliver course content, the faculty member chose to create a series of video clips in the following manner. During the initial offering of the course, she videotaped live lectures conducted with students in a campus-based distance education classroom. The Project Manager then created videoclips from the lecture and posted then within the online version of the course, making them synchronous with PowerPoint slides of key points. As these videoclips became outdated, the faculty members replaced them with new videoclips she created using an Apreso system which automatically synchronizes video to PowerPoint slides as the instructor presents content. The Project Manager posted these videoclips within the course to replace the outdated videoclips. The video- and audio-clips are brief (e.g., 1 min to 30 min), and the running time for each is posted to help students organize their viewing time. The content of each module was developed to equal 2 ½ hours of face-to-face class time.

Asynchronous Online Formats

Presentation of content. In its present incarnation, the Distance Education: Delivery course has the following components online within a Blackboard course management system format: (a) a videostreamed introduction of the instructor welcoming students to the course, (b) a course information section containing the syllabus and class schedule, (c) an announcement section for updating students on current distance education issues and course information, (d) 10 modules of course content presented through videostreaming synchronized with accompanying Powerpoint slides, (e) questions over course readings for each module that are posted as quizzes, (f) midterm and final project instructions, (g) discussion board questions over content for each module, (g) a "student café" discussion board, and (h) an online gradebook. In addition, an average of six readings per module from the professional literature on distance education can be accessed through the UK Distance Learning Library Services Website (also available in hard copy from an independent printing company). The management and support course has a similar format with the following exceptions: (a) PowerPoint slides are synchronized with audiostreaming rather than videostreaming, (b) links to distance education Websites are available and active, and (c) the delivery techniques videoclips have been deleted. Each course contains an optional module that consists of the videostreaming of a pre-taped and edited panel discussion between experienced distance education designers or administrators answering student questions on the course topic. These panel discussions were conducted in the UK television studio during the initial development of each course.

Interactions with learners. The courses are designed to facilitate three types of interactions during each module: (a) student-content, (b) student-instructor, and (c) student-student. Each of these types of interactions can take place online within the course management system or through personal e-mail.

Student-content interactions occur when the student views, listens to, and reads posted information and completes independent assignments. In the delivery course, this consists of developing a syllabus, course development timelines, guidelines for interactions, and the layout for a single lesson (e.g., graphic materials, content delivery, supplementary readings, assessment activities) within the student's discipline. In the course management system, this consists of identifying and posting articles and Website links relevant to each topic.

Students-instructor interactions occur through personal e-mail whenever students complete assignments. For example, students respond to the readings by answering essay questions posted by the instructors and receive individual feedback on their responses. In addition, students are encourage to e-mail the instructor at anytime throughout the courses and can expect to receive a response in a timely fashion. The fax machine and postal services are used as a backup system for interacting when Web-based technologies fail or are otherwise unavailable.

Student-student interactions occur through discussion questions posted for each topic. Students receive grades on their participation and are expected to both post a response to each discussion question and to respond to the postings of other students in the course. When necessary, the instructors join in these discussions to clarify points and give feedback on postings. In addition, students can interact informally with each other through the "student café" discussion board. While the instructors monitor all discussion board entries, they do not participate in the student discussion board.

FACTORS INFLUENCING DECISION TO USE ONLINE INSTRUCTION

Needs Addressed by Online Instruction

Distance education technologies are being used to deliver coursework in higher education across disciplines through a variety of modes, both synchronous (e.g., interactive video, satellite) and asynchronous (Web-based coursework). The professional literature has suggested the need for special skills for distance education delivery that include the ability to (a) plan course sessions using a variety of technologies, (b) work collaboratively with others, (c) facilitate interactions with distant students, (d) deliver course content in an effective manner, and (e) use accurate testing and evaluation strategies. While technologies are always subject to breaking down or becoming outdated, adequate preparation can alleviate many of the frustrations associated with distance education. Students often are willing to tolerate the nuances of technology if the instructor is competent and skilled in delivering the content. The purpose of the online courses in distance education at UK is to prepare distance educators who are skilled in delivery across technologies and who understand the issues involved in offering distance education programs in special education and other disciplines.

REASONS FOR SELECTION OF SPECIFIC ONLINE FORMATS

In selecting the delivery mode, the developers had three purposes: (a) to demonstrate the effective use of distance education technology while teaching students about its use, (b) to make the courses flexible for nontraditional students, and (c) to be able to reach students across a wide geographic area. Thus, they decided to develop the courses for Web-based delivery using different delivery techniques that included audioclips, videoclips, PowerPoint slides, and the Blackboard announcement, testing, and discussion functions.

OUTCOMES AND IMPLICATIONS OF ONLINE INSTRUCTION

IMPACT OF ONLINE INSTRUCTION ON THIS PROGRAM

The two online courses on distance education have been

offered annually at UK since the fall of 2000. To allow sufficient student-instructor interaction, course enrollment has been limited to no more than 12 students per semester. Students participating in the program have included UK and USU doctoral students in special education; UK graduate students from disciplines such as agriculture, communication, instructional design, law, and women's studies; and UK and KCTCS faculty from disciplines such as biology, foreign language, math, and sociology. Each of these students were either already involved or planning to be involved in distance education as part of their employment at institutions of higher learning. Formative and summative feedback from course evaluations have provided evidence that the students valued the course content and were satisfied with its delivery. In addition, a follow-up survey of the special education doctoral students involved in the initial delivery of the original course showed that those students were using what they had learned in their subsequent employment sites.

Opportunities/Advantages/Benefits of Online Instruction

The courses have been successful in a number of ways. First, the courses have provided students with a foundation for distance education by making them aware of delivery and administrative issues and giving them the opportunity to practice delivery techniques with feedback from an experienced distance educator. Second, the courses have provided students with exposure to numerous technologies and delivery issues across disciplines and levels of higher education. Third, the courses have allowed the students to form a network through interactions with other students also involved in distance education across a wide geographic range. Finally, the courses have provided students with a background in distance education that will increase their marketability in higher education. The online mode of delivery has allowed students to access course content as distance education students, making them more sensitive to and aware of issues that their own distance education students' experience. In addition, the asynchronous online mode of delivery has been a flexible mode of delivery for fulltime students and faculty to access at their convenience across time zones and geographic locations distant to UK.

Problems/Disadvantages/Limitations of Online Instruction

In spite of the success of the courses, the instructors have identified a number of challenging delivery issues that have required attention and subsequent changes in delivery methods. The instructors have shared these issues with students as they arise to demonstrate the need for flexibility and patience in online delivery. The following issues that are detailed in the following paragraphs represent only a few of the challenges that the instructors have faced in the delivery of their online courses.

First, both instructors have found that grading has taken more time than they originally allotted. Because the instructors respond individually to each student on each assignment, grading typically takes two full days per topic. The instructors have found that it is crucial to grade products immediately as they are submitted in order to be finished before students begin submitting products on the next topic. Although timelines are set for the students to progress through the course as a group, some products come in late due to problems beyond the students' control. In these cases, the policy is that all products submitted by the due date are graded and returned immediately, but late products are not graded until the instructor finds time. Another grading issue has occurred in the submission of assignments as attachments within the course management system or through person e-mail. Some formats used by students are not compatible with the computer system used by the instructors. In these cases, students are asked to resubmit in an alternate computer format or to submit as hard copies through fax or mail, causing delays in submission timelines.

Second, keeping timelines has been problematic. Students sometimes miss deadlines due to personal issues, such as illnesses, work conflicts, or technology problems (e.g., the UK server being down due to power failure or overuse or incompatible linkages when students are traveling). The instructors found that a submission deadline of 7:00 a.m. on Monday mornings was problematic because the majority of UK students often access online courses on Sunday night, thus overloading the system. The instructors require students to contact them immediately to explain the reason for missed deadlines; if excused, they are to turn the product in as soon as possible using an alternate mode of submission, if necessary.

Third, communication in online courses can be overwhelming. The instructors have found that students need to communicate almost daily on some aspect of the course and most often do this via e-mail. Thus, the instructors have found that the initial part of each day is spent responding to students' e-mails. When a student has asked a question of interest to all students, the instructors either have posted a response on the course management system bulletin board or sent out duplicate e-mail messages to all of the students. In some cases, the instructors have scheduled face-to-face or telephone appointments with students to discuss individual issues.

Fourth, due to the dynamic nature of distance education, the instructors are committed to continually updating the courses. This, however, requires time and funding on the part of both the instructors and the support staff. To keep revisions manageable and costs under control, the instructors have decided to fully update the courses at 4 to 5 yr intervals. In addition, they have added to the "shelf life" of each course by making posted audio and video content generic while addressing specific content through readings that are updated on an annual basis.

Fifth, understaffing at UK has caused the instructors to limit course enrollment per semester to 12 students. Since they anticipate that the demand for the courses will increase when the UK distance education certificate is finalized and as collaborative relationships with institutions within and outside of the state continue to develop, they are searching for potential instructors for additional sections.

Sixth, university technology upgrades and changes in course management systems have been problematic. During one semester, students found that they had to upgrade the version of Real Player on their computer mid-course in order to continue to access the videostreamed content of the course when UK upgraded their systems. When UK moved from a

TopClass to a Blackboard course management system, the entire content of the delivery course had to be redesigned and formatted within a short time frame.

Seventh, the most recent challenge to online course delivery has arisen with UK's mandate to make all courses accessible to students with disabilities in compliance with the Americans with Disabilities Act. Thus, the instructors are working on ways to supplement audio components with closed captioning, windows with manual signing, or text-based transcriptions, as well as ways to provide audio descriptions of video components. The university and the instructors have yet to reach consensus on how this endeavor will be funded.

Suggestions or Recommendations for Others Considering Online Instruction

The instructors offer the following guidelines for those who may be developing online courses, programs, or certificates to prepare distance educators.

First, it is never too soon to begin planning for the development of a Web-based course, especially if there will be a need to secure funding or the commitment of a development team. Second, course development can become expensive, especially if components, such as videostreaming or elaborate graphics, are employed, and sources of funding may come from internal or external sources that are time limited; thus, specific timelines are crucial in assuring that funds will be available as development activities are completed.

Third, video and audio should be used only when it adds to the content of the course in a way that facilitates learning because these components are expensive to develop and may make technology less accessible for learners; these components were used in these courses because the instructors wanted to give the students the experience of learning in this manner.

Fourth, the delivery of Web-based classes can be a major consumer of instructor time since the success of online learning is dependent on timely communication between instructors and students when they do not meet face-to-face; thus sufficient planning should be devoted to determining how and when communication will take place, communication time with students should be overestimated, students in an online course should be limited to a reasonable number, and instructors should take advantage of listservs, bulletin boards, and global announcement systems where messages can be posted, accessed, and read by the entire class.

Fifth, online instructors need to strike a balance in determining course assignments, making sure that they provide enough assignments to make distant students feel connected and provide sufficient and timely feedback on learning; in the absence of automatic feedback and pre-programmed grading systems, instructors need to determine the number and type of assignments per student that can be graded in a reasonable amount of time.

Sixth, one of the great truths about working with technology is that it will break down when it is least convenient, and overreacting in a hostile manner will not change this fact and may be a deterrent to recruiting new distance educators into the field; thus, online instructors should have a clear backup plan and initiate it with a positive attitude, saving their energy for getting the problem solved and hoping that their students on the receiving end will respond in kind.

Seventh, an online instructor should never limit course delivery to a single mode if it does not serve the purposes of the course and should consider options that include occasional face-to-face meetings conducted through audioconferencing or interactive video or chat rooms; in addition, the online instructor should consider the needs and abilities of each new group of students and act accordingly.

LIST OF RESOURCES

Books and Articles

Baird, M. (1998). Training distance education instructors. *Adult Learning, 7,* 24-26.

Blackhurst, A. E., Lahm, E. A., & Hales, R. M. (1998). Using an education server software system to delivery special education coursework via the World Wide Web. *Journal of Special Education Technology, 13*(4), 78-98.

Casey, C. (1999). Accessibility and the educational Web site. *Syllabus, 13*(2), 26-30.

Collins, B. C. (1997). Training Rural Educators in Kentucky through Distance Learning: A model with follow-up data. *Teacher Education and Special Education, 20,* 234-248.

Collins, B. C. (2001). Guidelines in distance learning content delivery. In B. L. Ludlow & F. Spooner (Eds.), *Distance education in special education: Personnel preparation applications.* Reston, VA: Council for Exceptional Children.

Collins, B. C., & Hess, J. M. (2000). In S. Smith, Teacher education: Associate editor's column. *Journal of Special Education Technology, 15*(4), 40-43. Available: http://jset.unlv.edu/15.4/asseds/

smith.html

Collins, B. C., Schuster, J. W., Ludlow, B. L., & Duff, M. (2002). Planning and delivery of online coursework in special education. *Teacher Education and Special Education, 25,* 171-186.

Gilbert, L., & Moore, D. R. (1998). Building interactivity into Web courses: Tools for social and instructional interaction. *Educational Technology, 38,* 29-35.

Harrison, N., & Bergen, C. (2000). Some design strategies for developing an online course, *Educational Technology, 40,* 57-60.

Hodgeson, P. (1999). How to teach in cyberspace. *Techniques: Making education and career connections, 74,* 34.

Keifer-O'Donnell, R., & Spooner, F. (2002). Effective pedagogy and e-learning. *Teacher Education and Special Education, 25,* 168-170.

Lawrence, B. H. (1998-97). Online course delivery: Issues of faculty development. *Journal of Educational Technology Systems, 25*(2), 127-131.

Ludlow, B. L. (2001). Technology and teacher education in special education: Disaster or deliverance? *Teacher Education and Special Education, 24,* 145-163.

Ludlow, B. L., & Brannon, S. a. (1999). Distance education programs preparing personnel for rural areas: Current practices, emerging trends, and future directions. *Rural Special Education Quarterly, 18*(3/4), 5-20.

Meyen, E. L., Tange, P., & Lian, C. H. T, (1999). Developing online instruction: Partnership between instructors and technical developers. *Journal of Special Education Technology, 14,* 18-31.

Spooner, F. (2001). Technology delivery without disaster. *Teacher Education and Special Education, 24,* 140-142.

Wisher, R. A., & Curnow, C. K. (1999). Perceptions and effects of image transmission during internet-based teaching. *The American Journal of Distance Education, 13,* 37-51.

Web Sites

http://www.uky.edu/DistanceLearning
http://www.uwex.edu/disted
http://www.usdla.org>http://www.usdla.org
http://www.ed.psu.edu/acsde/

Contact Information

Belva C. Collins, Ed.D.
Professor and MSD Program Faculty Chair
Department of Special Education and Rehabilitation Counseling
229 Taylor Education Building
University of Kentucky
Lexington, KY 40506-0001
859-257-8591
bcoll01@uky.edu

Constance M. Baird, M.Ed., MLIS
Distance Learning Programs
2-2 W. T. Young Library
University of Kentucky
Lexington, KY 40506
859-257-8135
bairdc@uky.edu

DISTANCE EDUCATION: SOME CONCLUDING THOUGHTS

CLEBORNE D. MADDUX
University of Nevada, Reno

For those of us who are interested in technology and excited by technological innovation, the last 25 years have been a great quarter century to be alive – perhaps the greatest in history. After all, it could be argued that the pace of technological change, as well as its effect on worldwide culture, has never been greater. One of my favorite statistics in this regard is that, when President Clinton took office for his first term, there were only about 50 pages on the Web. Today, no one knows for sure how many pages reside there, but one of the popular search engines recently claimed to be indexing 19.2 billion Web pages! That claim is somewhat controversial and may or may not be inflated, but, regardless of exactly how many billions of pages are actually out there, it is clear the Web has undergone unprecedented growth, and this phenomenal growth continues today.

It was inevitable that rapid and profound technological

changes would result in equally rapid and profound changes in nearly every walk of life. We have so far seen only the beginning of these changes as computers and information technology in general have become ubiquitous in our daily lives. Additionally, as the articles in this volume aptly demonstrate, dramatic changes with their roots in the technology revolution have now begun to find their way into higher education in the United States. These changes are especially apparent in the area of distance education, but they are probably only the forerunners of even more startling changes to come in higher education in general.

I am not one of those educators who predict the demise of universities or the disappearance of face-to-face education. I believe there always will be large numbers of undergraduates who value a campus experience, and I think it unlikely that the majority of doctoral students will ever be satisfied with a program that offers anything less than total immersion in the subculture of their disciplines, a goal unlikely to be achieved soon in a virtual environment.

However, I think it would not be an exaggeration to say that higher education in this country, at least at the master's level, is poised on the brink of a distance education revolution. I think it possible that one day the majority of master's degree students in most programs at most institutions will complete most or all of their coursework totally online.

That trend is already beginning, as documented by the authors of the various chapters in this volume, many of whom report a significant increase in students when their programs became available online. Further, as I talk to colleagues across the country, many of them professors in information technology in education programs, most report a steep decline in the number of students in their face-to-face master's degree programs. Those with online programs are experiencing an increase in those programs, while many of those with no online option say that the decline is due to students opting for online programs offered by a rapidly growing number of legitimate and not-so-legitimate institutions.

I believe that this trend will rapidly find its way into other master's programs – those across colleges of education, as well as those across most other disciplines. The fact that the trend seems to have taken place first in information technology in education and in special education is understandable. Students interested in technology are naturally drawn to the idea of an online course or an online program, while the shortage of special education teachers, particularly in remote areas, provides strong motivation to accept the idea and the option of online certification and master's degree programs.

It is interesting to observe how different institutions are responding to this new trend. Some, like the institutions of the authors of articles in this volume, are responding carefully but positively and are creating online courses and programs that meet the needs of this new constituency. At the other end of the response continuum are others that seem to be taking the position that ignoring the emerging demand for online study will make it go away. I believe that is unlikely. Like small computers themselves and like the Internet and the World Wide Web, distance education now has gained so much cultural momentum that we probably could not prevent further wide public acceptance and significant future growth even if we set out to do so.

All of this begs the question of whether or not the trend toward distance education is a good thing. That, of course, depends on a number of variables, not least of which is the quality of our response. Quality control is certainly a major consideration and is the only thing that can separate our efforts from those of the well known "diploma mills," which, not surprisingly, are rushing to meet the burgeoning demand for online courses and programs.

I am only a little concerned about quality control, because most legitimate institutions seem to be doing a good job of carefully considering which courses are and are not suited to online delivery and are taking care to use the results of program evaluation to modify their offerings. This is certainly true of the programs summarized in this book.

My only concern is related to the difference in the professional subculture of academic departments and that of the campus administrative units responsible for distance education. The entrepreneurial spirit and economic imperative of the latter sometimes runs counter to concerns of faculty about academic integrity. The best solutions I have seen involve a system of checks and balances that involve the personnel in both units in a formal process of decision-making about what will be offered and how it will be offered. The trick, of course, is to set up the process in such a way that those in distance education units do not focus too narrowly on the economic bottom line and those in academic departments do not act simply as independent contractors who are not accountable to peers or any academic authority. As always, the goal is true collaboration of the two units while avoiding the generation of so much red tape that nothing can be accomplished in a timely manner. As we all know, universities are not well known for their ability to create such systems.

Earlier, I suggested that we are poised on the brink of a distance education revolution. I believe the only thing standing in the way of that revolution is bandwidth and that the necessary increase in bandwidth is imminent. It will be slower to reach rural and remote areas, but widely available fast connections, such as DSL lines, as well as the eventual availability of Internet II are developments that are on the horizon. When they arrive, high quality voice and video over the Internet will be practical and will begin to bring many of the advantages of face-to-face education to distance learners.

University special educators have long been leaders in innovative program delivery, including distance education. That has been especially true of those institutions serving large rural areas. Perhaps that accounts for the fact that, as I read through the individual contributions to this volume, I am struck by the presence of certain commonalities. There are trends that are documented by nearly every included program summary, and it is remarkable the degree to which those trends mirror trends in higher education distance education across disciplines and program areas. The following section is a partial list and brief discussion of such trends in no particular logical order.

1. **Programs are moving rapidly to the Internet and the Web as the main vehicle for course delivery.**

There are still some programs making use of specialized equipment used to conduct real time programs in distance education "centers," but the trend is definitely toward the use of the Web. It is clear that the development and popularization of the Internet and the Web are responsible for the current resurgence of interest in distance education. Software programs, such as WebCT and Blackboard, are common choices as platforms for online delivery.

2. **Course and program offerings are increasing rapidly.**

Kay Bull's report that offerings in his institution grew from 10 in 1997 to over 1300 in 2005 are probably at the extreme end of the growth continuum, but rapid growth is occurring almost everywhere and across many disciplines.

3. **Institutions are moving quickly from offering individual courses online to offering entire online programs and from involving students in their immediate areas to including students in other states and other countries.**

Barbara Ludlow mentioned the movement from individual courses to entire programs with reference to programs in low incidence disabilities, but the same trend is occurring nationally across disciplines. Harvey Rude and Kay Ferrell identified target students as those living in the entire rural and remote locations of the Mountain West as well as those in other national and international locations where such

4. STUDENTS WANT TO HAVE THE OPTION OF ONLINE COURSES AND PROGRAMS.

Support for this contention is found in nearly every chapter of this book. Even those students who say they prefer face-to-face education say they want to have the online option because of its convenience and flexibility. Belva Collins does a good job of summarizing some of the research on student preferences. Additional evidence of student demand is the typical growth in number of students in programs when an online option is developed. Harvey Rude and Kay Ferrell report that enrollments in Rude's institution grew from approximately 80 students in four traditional programs to 277 students in the same programs after institution of online delivery. Sue Steinweg and Sandra Warren assert that the online program increased student enrollment to the point that additional faculty and resources were obtained.

5. DESIGNING, TEACHING, AND UPDATING ONLINE COURSES IS EXTREMELY TIME-CONSUMING.

Nearly every chapter in this book makes the point that online courses require more time and more work than equivalent face-to-face courses. That fact has been emphasized again and again in the general literature on online education.

6. **University Information Technology (IT) units and faculty teaching online courses need more and better communication.**

Belva Collins and Constance Baird make the point that technology upgrades and changes in course management systems cause problems for online instructors. David Forbush and Robert Morgan agree and emphasize the importance of advanced instructor notification of technical changes that may impair instructional delivery. In the past, IT staff on most campuses have been free to make whatever changes in technology they cared to make without prior notification outside their own departments. With the advent of distance education courses, however, even something as minor as changing the URL of a university Web site can cause major problems for online faculty and students and should take place only after consultation and plenty of prior notification. This will require a new spirit of collaboration on the part of IT staff. At the same time, faculty need to recognize that IT staff usually are not educators, and be patient when hardware and software problems do occur.

7. **While there are concerns about social isolation or lack of participation in discussions in online classes, many students are motivated to communicate more in online classes than in face-to-face classes.**

Betty Epanchin, Elizabeth Doone, and Karen Colucci comment that online discussions involve everyone and cannot be dominated by the most vocal students, as they sometimes are in face-to-face classes. They also found that online courses stimulated more timely communication and feedback and that the online format has a tendency to break down inhibitions to candidly express opinions and concerns. While we often hear most about potential communication problems in online education, it is clear that there are certain advantages to the online environment when it comes to student participation.

8. **THERE IS A TREND FOR ONLINE COMMUNICATION AND DISCUSSION TO BE CONDUCTED PRIMARILY THROUGH ASYNCHRONOUS RATHER THAN SYNCHRONOUS MEANS.**

Dennis Knapczyk mentions that the last few years at his institution has involved a movement from videoconferencing to online format with asynchronous class discussions. The reasons he cites are that non-real time discussions, such as bulletin boards, give teachers and students more time to prepare thoughtful contributions and makes it possible for a discussion to extend over several days and involve those who could not be present for a real time discussion due to scheduling demands. I suspect this trend may not continue, or will slow once solutions to the bandwidth problem make streaming video and Web-based conferencing software more practical for use by all students. Even then, however, the advantages of asynchronous discussion and student demand for scheduling flexibility will probably mean that most courses will eventually make use of both types of communication tools.

9. **There is a strong motivation to develop online programs to serve rural areas and in disciplines in which there are shortages and high attrition rates.**

Barbara Ludlow and Michael Duff refer to this trend with regard to the Appalachian region. Martin Agran refers to the importance of the shortage of qualified personnel, especially in the area of severe disabilities, and the trend is mentioned by many of the other authors in this book. The same trend is present nationally in the other helping professions and even extends into fields such as engineering and medicine.

10. **Good technical support is essential for both faculty and students.**

Many of the authors in this book make that point. Katherine Mitchem adds that it is important that participants in distance education can obtain support beyond the regular business day – preferably 24 hours a day. Joan Sebastian, Stuart Schwartz, and Jane Duckett also emphasize the need for support that is available 7 days a week, and report that their program makes use of support provided by email and toll free phone. If students have to wait for the next business day to obtain technical help, the flexibility they value so much in online education is badly compromised. This is a widespread problem, since help desks at many universities keep very limited hours.

11. **Distance Education presents many opportunities for research.**

Jane Pemberton, Tandra Tyler-Wood, and L. Nan Restine

report the results of some of their research generated in conjunction with their online classes. Joan Sebastian also has been involved in research activities, as have several other authors in this book. The potential for research needs to be emphasized to faculty who complain that online classes are so demanding that they do not have time to do the research they need for promotion and tenure. In addition, there is a critical need for research into many aspects of distance and online education.

12. **THERE IS A NEED FOR ADMINISTRATIVE SUPPORT FOR DISTANCE EDUCATION LOCATED AT THE COLLEGE OR DEPARTMENTAL LEVEL.**

Teresa Rowlison reports her institution has appointed an assistant director for distance education for the college of education. Joan Sebastian, Stuart Schwartz, and Jane Duckett recommend a departmental coordinator be appointed. As more online courses are offered, the need for such an administrator will likely continue to increase. Liaison with staff in distance education units as well as overall planning and scheduling is sure to become more problematic and require administrative attention. The rapidly increasing importance of program assessment also will require that someone assume responsibility for development and application of an assessment system specifically tailored for online offerings and for a process by which results are used to modify the program.

13. **ONLINE COURSES SHOULD NOT ENROLL LARGE NUMBERS OF STUDENTS IN A SINGLE SECTION.**

Although the optimum number of students in a course depends on many factors, including the content of the course itself and whether or not a graduate assistant is provided, most experienced instructors recommend somewhere between 10 and 25 students. Teresa Rowlison reports her institution restricts enrollment to 20 in online classes.

THE FUTURE OF DISTANCE EDUCATION

I am always hesitant to make predictions about the future of technology, in or out of education. That is because I have witnessed the way reality has a way of eclipsing what seem the most extravagant and unlikely predictions. Nevertheless, I will venture a few very modest suggestions about what I think the future may hold.

I already have mentioned that I think most future master's degrees will be obtained in online programs and that there will be almost unlimited bandwidth available to most instructors and students. Vastly increased bandwidth will make it possible and practical to use videoconferencing software that will enable us to see and hear all our students and for them to see and hear us, all in real time. I think increasing numbers of university faculty will have their entire teaching load consist of totally online courses. I believe that traditional campuses and traditional face-to-face courses and programs will continue to exist, but the growth in online education, particularly at the master's level but also at every level, will mean that most institutions will not continue to grow enrollments in their traditional programs. Most of the enrollment growth will be online growth. University IT departments will experience great growth as institutions are forced to increase their investment in hardware and software and also in technical support.

Production of multimedia material for use in online courses will become an important activity for IT staff. Colleges of extended studies also will experience a great deal of growth as more and more courses and programs go online. Accrediting agencies will begin to devote a great deal more attention to the development of standards for online courses and programs. Computers will continue to get smaller and less expensive, and I believe that mobile technology will become much more important in education than it currently is. Wireless networks and proliferation of wireless hot spots, many covering entire cities or regions, will facilitate the growth in importance of all kinds of wireless devices.

CONCLUSIONS

I began this final chapter by suggesting that, for those of us who like technology, the last 25 years may have been the best of all times to be alive. I suspect that it will not be remembered that way, however, because there is every indication that the next 25 years will be even more interesting and even more exciting. The future of distance education in general and online education, in particular, seems secure, and the authors who have described their programs in this book can be proud of the fact that they have participated in the pioneering work of a field that already has begun to transform American higher education.

ISBN 1412082389

Made in the USA
Lexington, KY
29 July 2011